CW01642562

THE NUCLEAR QUESTION IN THE MIDDLE EAST

MEHRAN KAMRAVA

Editor

The Nuclear Question in the Middle East

CIRS
CENTER FOR INTERNATIONAL AND REGIONAL STUDIES

GEORGETOWN UNIVERSITY
SCHOOL OF FOREIGN SERVICE IN QATAR

Published in Collaboration with
Georgetown University's Center for International and Regional Studies,
School of Foreign Service in Qatar

HURST & COMPANY, LONDON

Published in Collaboration with
Georgetown University's
Center for International and Regional Studies,
School of Foreign Service in Qatar

First published in the United Kingdom in 2012 by
C. Hurst & Co. (Publishers) Ltd.,
41 Great Russell Street, London, WC1B 3PL
© Mehran Kamrava and the Contributors, 2012
All rights reserved.
Printed in the United Kingdom

The right of Mehran Kamrava and the Contributors to be identified as the authors of this publication is asserted by them in accordance with the Copyright, Designs and Patents Act, 1988.

A Cataloguing-in-Publication data record for this book is available from the British Library.

ISBN: 978-1-84904-211-6 paperback

This book is printed using paper from registered sustainable and managed sources.

www.hurstpub.co.uk

Contents

Acknowledgments

The chapters in this volume grew out of two working group meetings held under the auspices of the Center for International and Regional Studies of Georgetown University's School of Foreign Service in Qatar. Grateful acknowledgment goes to all the participants in the working groups for their intellectual contributions to the project that eventually resulted in this book. Along the way, invaluable assistance also came from the capable and professional staff of the Center for International and Regional Studies, namely Zahra Babar, John Crist, Suzi Mirgani, Kasia Rada, Naila Sherman, Nadia Talpur, and Maha Uraidi. I gratefully acknowledge their many contributions to the working groups and to this volume. Perhaps the greatest sources of support and inspiration have been my wife Melisa and our daughters Kendra and Dilara. It is to them that I dedicate this book.

About the Contributors

Gawdat Bahgat is Professor at the Near East South Asia Center for Strategic Studies at the National Defense University. In addition to a number of journal articles, his books include *Proliferation of Nuclear Weapons in the Middle East, The Persian Gulf at the Dawn of the New Millennium, American Oil Diplomacy in the Persian Gulf and the Caspian Sea* and *Israel and the Persian Gulf: Retrospect and Prospect.*

Kayhan Barzegar is a faculty member at the Science and Research Branch of the Islamic Azad University, and Director of the Institute for Middle East Studies in Tehran. He is also an Associate at Harvard Kennedy School's Belfer Center for Science and International Affairs. He is the author, most recently, of *Iran's Foreign Policy in the New Iraq, Iran's Foreign Policy Strategy after Saddam, Balance of Power in the Persian Gulf* and *The Paradox of Iran's Nuclear Consensus.*

Avner Cohen is Senior Fellow at Center for Non-proliferation Studies and Professor in the Graduate School of International Policy and Management at the Monterey Institute of International Studies. He was previously a Public Policy Scholar at the Woodrow Wilson International Center for Scholars. A two-time winner of MacArthur Foundation research and writing awards, in addition to a number of journal articles and edited volumes, his books include *The Worst Kept Secret: Israel's Bargain with the Bomb and Israel and the Bomb.*

Mehran Kamrava is Professor and Director of the Center for International and Regional Studies at Georgetown University in Qatar. He is the author, most recently, of *Iran's Intellectual Revolution* and *The Modern Middle East: A Politi-*

cal History Since the First World War. His edited volumes include *The International Politics of the Persian Gulf* and *Innovation in Islam: Traditions and Contributions*.

Mustafa Kibaroglu is the Chair of the Department of International Relations at Okan University in Istanbul. In addition to numerous articles in journals such as *Security Dialogue, Nonproliferation Review, Bulletin of the Atomic Scientists, Middle East Journal, Middle East Policy, Middle Eastern Studies, European Security* and *Arms Control Today*, Kibaroglu is the author of *Global Security Watch—Turkey: A Reference Handbook* and editor of *Defense Against Weapons of Mass Destruction Terrorism*.

Thomas W. Lippman, former Middle East bureau chief of *The Washington Post*, is an adjunct Senior Fellow of the Council on Foreign Relations and an adjunct scholar at the Middle East Institute. He has been studying and writing about affairs in the Middle East for thirty-five years. In addition to a number of journal articles, he is the author, most recently, of *Arabia on the Edge: The Perilous Future of an American Ally*, as well as *Inside The Mirage: America's Fragile Partnership with Saudi Arabia, Egypt After Nasser: Sadat, Peace, and the Mirage of Prosperity*, and others.

Giacomo Luciani is Scientific Director of the Master in International Energy, Paris School of International Affairs at the Sciences-Po, and a Princeton University Global Scholar; he is also a visiting professor at the Graduate Institute of International and Development Studies and co-Director of the Executive Master in International Oil and Gas Leadership. He is the Team Leader in an EU Commission sponsored project to establish a GCC-EU Clean Energy Network. He has published extensively on the political economy of the GCC countries and global energy issues.

Mari Luomi is the 2011–12 Post-Doctoral Fellow at the Center for International and Regional Studies of Georgetown University in Qatar. In addition to various edited and single-authored publications for the Finnish Institute of International Affairs, her most recent journal article "Abu Dhabi's Alternative-Energy Initiatives: Seizing Climate-Change Opportunities" was published in *Middle East Policy*, 16, 4 (Winter 2009).

Maria Rost Rublee is Senior Lecturer in International Relations at the Australian National University. She is the author of *Nonproliferation Norms: Why States Choose Nuclear Restraint*, winner of the 2010 Alexander George Book

Award for best book in political psychology. Rublee is an editor for *International Studies Perspectives*, and has published in journals such as *International Studies Review*, the *Nonproliferation Review* and *Comparative Political Studies*.

Etel Solingen is Chancellor's Professor at the University of California, Irvine. Her most recent book, *Nuclear Logics: Contrasting Paths in East Asia and the Middle East*, was awarded the American Political Science Association's Woodrow Wilson Foundation Award and the Robert Jervis and Paul Schroeder Award.

1

The Nuclear Question in the Middle East

Context, Complexities, and Paradoxes

Mehran Kamrava

Much of the scholarly and policy attention concerning nuclear programs, both civilian and military, revolves around questions of proliferation and non-proliferation. Why and when do states choose to proliferate or abstain from proliferation? What are the causal domestic and international dynamics at work? What potential linkages are possible between civilian and military nuclear programs? And, what are the precise interplays of individual decisions and institutional forces that result in certain proliferation or non-proliferation outcomes? These and other similar questions have occupied prominent positions in much of the contemporary scholarship on nuclear studies.[1]

[1] In addition to the literature cited below, a small sampling of this scholarship includes Michael Brown (ed.), *Going Nuclear: Nuclear Proliferation and International Security in the 21st Century* (Cambridge, MA: MIT Press, 2010); Jacques E.C. Haymans, *The Psychology of Nuclear Proliferation: Identity, Emotions and Foreign Policy* (Cambridge University Press, 2006); Sverre Lofgaard and Bremer Maerl (eds), *Nuclear Proliferation and International Security* (London: Routledge, 2007); Robert Rauchhaus, Matthew Kroenig and Erik Garzke (eds), *Causes and Consequences of Nuclear Prolif-*

Concurrent with, and reinforcing, the focus on proliferation has been a rediscovery over the last decade or so of the virtues of nuclear power as a relatively reliable and environmentally friendly energy source. Particularly until the devastating March 2011 earthquake in Japan, which quickly led to a series of explosions at some of the affected nuclear power plants and leakages of radioactive materials into the air and the seawater, the world was witnessing a "nuclear renaissance" that reached into the four corners of the globe.[2] The Middle East was no exception, with Arab League states deciding in 2007 to expand the peaceful use of nuclear technology "for all fields of sustainable development, with due consideration of the diversity of their needs."[3]

Given this rediscovery of the virtues of nuclear power around the globe, it is important to differentiate between two related phenomena: a potential *growth* in the production of nuclear energy by states that already have nuclear power production facilities, and the potential *spread* of nuclear power plants and related facilities to states that are new entrants to the "nuclear energy club."[4] To alleviate potential concerns about proliferation, multilateral nuclear arrangements have grown in popularity as a way to strengthen the non-proliferation regime through the "de-nationalization" of sensitive fuel cycle facilities in non-nuclear weapons states (NNWSs).[5] Access to nuclear energy technology by non-nuclear weapons states inevitably leads to concerns about the possible acquisition of nuclear weapons, as both require similar skills and technologies, the most notable being uranium enrichment and plutonium reprocessing capabilities.[6] There are two key ingredients for making nuclear weapons: separated plutonium, and highly enriched uranium (HEU). Civilian nuclear power plants (operated by civilians) produce low enriched ura-

eration (London: Routledge, 2011); and James A. Russell, "A Tipping Point Realized? Nuclear Proliferation in the Persian Gulf and Middle East," *Contemporary Security Policy*, Vol. 29, No. 3 (December 2008), pp. 521–37.

2 Spencer W. Semes, "A Nuclear Renaissance," *IEEE Power and Energy Magazine* (November–December 2006), pp. 34–42.

3 Quoted in Mohamed I. Shaker, "Nuclear Power in the Arab World and the Regionalization of the Nuclear Fuel Cycle: An Egyptian Perspective," *Daedalus*, Vol. 139, No. 1 (Winter 2010), p. 97.

4 Steven E. Miller and Scott D. Sagan, "Nuclear Power without Nuclear Proliferation?" *Daedalus*, Vol. 138, No. 4 (Fall 2009), p. 9.

5 Bruno Pellaud, "Multilateral Nuclear Arrangements: Status and Outlook," *Esarda Bulletin*, No. 36 (June 2007), p. 11.

6 Miller and Sagan, "Nuclear Power without Nuclear Proliferation?", p. 13.

nium (LEU) and give countries access to plutonium separated for civilian purposes. But the potential for converting LEU to HEU can significantly increase the possibility of latent proliferation and "breakout" capacity, whereby fissile materials for weapons purposes are swiftly, and perhaps even openly, produced.[7]

Proliferation concerns abound. At the same time, there is genuine frustration by the non-nuclear states when the nuclear weapons states pressure them to observe the non-proliferation regime that they themselves are unwilling to abide by.[8] Mohamed ElBaradei, the former head of the International Atomic Energy Agency, sums up the consequences of this "double standard" as follows:

> The most fundamental problem with the nuclear nonproliferation regime is, in itself, a double standard: the inherent asymmetry, or inequality, between the nuclear haves and have-nots, exacerbated by the continuing reliance by the nuclear power states on nuclear weapons and their lack of progress on nuclear disarmament. Worst, rather than moving to fulfill their commitments to disarm, most have modernized their arsenals and continue to develop new types of weapons. For countries that do not have such weapons and do not fall under a nuclear umbrella protection arrangement such as NATO, this reinforces the perception that the acquisition of nuclear weapons is a sure path to power and prestige, an insurance policy against attack.[9]

Articles IV and VI of the Non-Proliferation Treaty (NPT), which define rights and obligations with respect to peaceful use of nuclear energy and nuclear weapons disarmament respectively, are often seen as particularly preferential to supplier, nuclear-weapons states.[10] For their part, given the opportunity, most Middle Eastern states could either enter into joint partnerships for the construction of, or purchase outright, relatively advanced nuclear power plants. According to one observer, "it is clear that Arab states would have the expertise, uranium ore deposits, research reactors, fuel fabrication skills on a small scale, accelerators, and other nuclear-related laboratories, including hot cell laboratories, necessary to develop an Arab nuclear fuel cycle."[11] Mutual

[7] Robert H. Socolow and Alexander Glaser, "Balancing Risks: Nuclear Energy and Climate Change," *Daedalus*, Vol. 138, No. 4 (Fall 2009), p. 36.

[8] Pellaud, "Multilateral Nuclear Arrangements," p. 11.

[9] Mohamed ElBaradei, *The Age of Deception: Nuclear Diplomacy in Treacherous Times* (New York: Metropolitan Books, 2011), p. 236.

[10] Robert H. Socolow and Alexander Glaser, "Balancing Risks: Nuclear Energy and Climate Change," p. 39.

[11] Shaker, "Nuclear Power in the Arab World and the Regionalization of the Nuclear Fuel Cycle," p. 98.

distrust and outright competition, as the case of the Gulf Cooperation Council has demonstrated, make the possible emergence of such joint partnerships in the Persian Gulf seem unlikely. Even more consequential is current distribution of power in the international community, and particularly the roles currently played by the United States and the European Union, making it highly unlikely for cooperative nuclear arrangements to emerge within the GCC or other Middle Eastern states because of potential breakout and proliferation possibilities.

The proliferation puzzle

As already noted, nuclear proliferation has been the subject of a number of thorough and incisive studies.[12] The primary drivers of nuclear proliferation may be broadly divided into the four categories of economic, strategic, institutional and normative. These drivers have been studied from theoretical perspectives that may be broadly categorized as realist and neorealist, institutionalist and constructivist.[13] Insofar as the actual causes of nuclear proliferation (or of its opposite, nuclear restraint) are concerned, as the case studies in this volume amply demonstrate, each case needs to be studied within its own specific, individual context. Proliferation, as we shall see shortly, can occur for more than one reason.

One of the more pervasive arguments about the causes of nuclear proliferation points to strategic concerns as a primary driver. Matthew Kroenig, for example, looks at the nature of the alliance and power relationships that exist between potential nuclear supplier and recipient states. More powerful states, he maintains, are less likely to provide sensitive nuclear assistance to less powerful states, as that would undermine their own strategic advantages in the international arena.[14] Along the same lines, states that are dependent on a

[12] Just two of the more recent studies in this category, which examine the issue both theoretically and empirically, include William C. Potter and Gaukhar Mukhatzhanova (eds), *Forecasting Nuclear Proliferation in the 21st Century: The Role of Theory, Volume 1* (Stanford University Press, 2010); and William C. Potter and Gaukhar Mukhatzhanova (eds), *Forecasting Nuclear Proliferation in the 21st Century: A Comparative Perspective, Volume 2* (Stanford University Press, 2010).

[13] William C. Potter and Gaukhar Mukhatzhanova, "Forecasting Proliferation: The Role of Theory, an Introduction" in Potter and Mukhatzhanova (eds), *Forecasting Nuclear Proliferation in the 21st Century: The Role of Theory, Volume 1*, pp. 1–12.

[14] Matthew Kroenig, "Exporting the Bomb: Why States Provide Sensitive Nuclear

superpower patron will also be less likely to provide sensitive nuclear assistance to other states.[15] Nevertheless, a state is more likely to provide nuclear assistance to another state with which it shares a common enemy.[16]

Apart from the types of alliances in the international arena and their influence on access to nuclear assistance, the nature and extent of proliferation are influenced more by context and specific circumstances than by any other single factor. This argument has been most persuasively articulated by Scott Sagan, who calls for a nuanced approach to the strategic causes of proliferation. Contextualizing the "proliferation begets proliferation" thesis, which informs the underlying assumptions of realist and neorealist perspectives,[17] Sagan maintains that "nuclear weapons proliferation and nuclear restraint have occurred in the past, and can occur in the future, for more than one reason: different historical reasons are best explained by different causal models."[18] In some instances, Sagan argues, states proliferate because of security concerns in order to safeguard their national interests. When alternative means of defense are insufficient to meet threats, nuclear proliferation may appear as an attractive option to pursue.[19] At other times, however, proliferation may be driven by domestic political actors, such as the state's nuclear energy establishment (such as technocrats, scientists, technical experts and the like), influential units within the armed forces, and individual politicians seeking to tap into popular and nationalist sentiments.[20] In still other instances, a state's actions in general and specifically in regards to proliferation may be guided by deeper norms and beliefs about what actions are legitimate in the international arena and what actions are likely to enhance its identity and prestige.[21]

"Nuclear hedging" is another potential driver of proliferation. Nuclear hedging consists of continuing efforts in nuclear research and development

---

Assistance," American Political Science Review, Vol. 103, No. 1 (February 2009), p. 127.

[15] Ibid., p. 117.

[16] Ibid., p. 116.

[17] Potter and Mukhatzhanova, "Forecasting Proliferation: The Role of Theory, an Introduction" , pp. 3–4.

[18] Scott D. Sagan, "Why Do States Build Nuclear Weapons?", *International Security*, Vol. 21, No. 3 (Winter 1996–7), p. 85.

[19] Ibid., p. 54.

[20] Ibid., pp. 63–4.

[21] Ibid., p. 73.

while abstaining from acquiring the weapon itself. It refers to "a national strategy of maintaining, or at least appearing to maintain, a viable option for the rapid acquisition of nuclear weapons, based on an indigenous technical capacity to produce them within a relatively short time frame ranging from several weeks to a few years."[22] Hedging enables national leaders to acquire nuclear knowledge and infrastructures while still remaining within the limits of the NPT regime. Kayhan Barzagar and Maria Rost Rublee examine the employment of nuclear hedging by Iran and Egypt—under quite different circumstances and contexts—in their respective chapters in this volume.

One of the aspects of nuclear proliferation that is often overlooked in studies on the topic is the contributing role of nuclear-exporting countries and the economics of proliferation. In recent years, a small but growing number of scholars have drawn attention to the importance of the export priorities of nuclear suppliers. Proliferation, they argue, is a result of a convergence of the economic interests of nuclear exporters and the strategic and energy priorities of importers. Although environmental concerns are not unimportant,[23] they are often secondary to economic considerations, especially insofar as nuclear-exporting countries are concerned. During the tenure in office of French President Nicolas Sarkozy, for example, France entered into a number of nuclear cooperation agreements (NCAs); it signed multiple NCAs with interested partners, including Saudi Arabia, Algeria, Jordan, Libya, Qatar, the United Arab Emirates (UAE) and Vietnam.[24] According to Peter Stoett, these NCAs have multiplied due largely to their lucrative nature, having become tools for legitimizing what has become an "export-oriented market."[25] Within the exporting countries, the impetus to export civilian nuclear technology is often a result of synergistic efforts by interest groups, governmental sub-units (such as commissions and government-employed technical experts), and, most notably, multinational corporations and other players in the export industry (for example Westinghouse and General Electric Corporations).[26]

[22] Ariel E. Levite, "Never Say Never Again: Nuclear Reversal Revisited," *International Security*, Vol. 27, No. 3 (Winter 2002–3), p. 69.

[23] Peter Stoett, "Toward Renewed Legitimacy: Nuclear Power, Global Warming, and Security," *Global Environmental Politics*, Vol. 3, No. 1 (February 2003), p. 114.

[24] Matthew Fuhrmann, "Spreading Temptation: Proliferation and Peaceful Nuclear Cooperation Agreements," *International Security*, Vol. 34, No. 1 (Summer 2009), p. 7.

[25] Stoett, "Toward Renewed Legitimacy: Nuclear Power, Global Warming, and Security," p. 99.

[26] Ibid., pp. 102–3.

These agreements often feature stringent safeguards meant to prevent possibilities of weapons proliferation (see Mari Luomi's chapter on the United Arab Emirates in this volume). Such anti-proliferation measures were further strengthened in 1994 by the informal Nuclear Supply Group, a coalition of nuclear suppliers that coalesced in the 1970s and has since sought to coordinate, and ostensibly to control, the supply and spread of nuclear materials to non-nuclear states.[27]

Despite these safeguards, Matthew Fuhrmann argues that NCAs invariably increase the chances of nuclear proliferation.[28] Nuclear agreements often reduce the expected costs of nuclearization campaigns and give leaders greater confidence that they can indeed produce a bomb. If the country's security environment worsens, then the chances of diverting a civilian program to a military one increase substantially. Peaceful cooperation, Fuhrmann maintains, is "consistently salient" in explaining both the start of a nuclear weapons program and the acquisition of weapons.[29] Fuhrmann sees a direct correlation between civilian nuclear assistance, a deteriorated security environment, and proliferation.[30] "Countries are highly unlikely to begin weapons programs in the absence of (nuclear) assistance—even if they face security threats. But if threats are present and states receive additional atomic assistance, the likelihood of beginning a bomb program spikes dramatically."[31] By increasing knowledge in nuclear-related matters and reducing the expected costs of weapons programs, "all forms of atomic assistance—whether it involves training scientists, supplying reactors, or building fuel fabrication facilities—raise the likelihood that nuclear weapons will spread."[32] Ultimately, he argues, "there is no such thing as 'proliferation-proof' atomic assistance."[33]

Fuhrmann's arguments are countered by Jennifer Morstein and Wayne Perry, who maintain that "nations seeking nuclear weapons trade fundamentally differently on the open market than do nations with solely energy intentions."[34] Nuclear aspirants create trade networks that tend to have three

[27] Ian Anthony, Christer Ahlstrom and Vitaly Fedchenko, *Reforming Nuclear Export Controls: The Future of the Nuclear Supply Group* (Oxford University Press, 2007), p. 21.

[28] Fuhrmann, "Spreading Temptation," pp. 8–9.

[29] Ibid., p. 9.

[30] Ibid., p. 35.

[31] Ibid., p. 33.

[32] Ibid., pp. 12–3.

[33] Ibid., p. 39.

[34] Jennifer Hunt Morstein and Wayne D. Perry, "Commercial Nuclear Trading Net-

distinctive features: components are procured through frequent, small purchases; what cannot be purchased legally is often acquired through illicit means such as smuggling or going through middlemen; and the larger, legitimate nuclear energy trade is often used to obscure weapons procurement efforts.[35]

The chapters by Etel Solingen and Maria Rost Rublee in this volume highlight two important contributions to the proliferation debate. Solingen looks at the overall position of a state within the international arena, and state leaders' perceptions toward greater international integration, as a broad—though not deterministic—factor in whether or not a state is likely to choose nuclearization. Rublee points to the importance of agency—bringing the individual back in, as she puts it—and of the social-psychological factors that are likely to prompt leaders to adopt certain decisions over others.

It is worth asking a question that is seldom asked: is nuclear proliferation indeed a catalyst for greater international tensions and instability? There is general consensus in policy circles, and also among most scholars, that the proliferation of nuclear weapons indeed increases international instability, crisis, and the risks of accidental nuclear detonation.[36] The arguments of these "proliferation pessimists" are countered by "proliferation optimists," who maintain that the spread of nuclear weapons actually enhances international stability as it tends to deter military aggression by more powerful states.[37] In fact, under certain circumstances, the spread of nuclear weapons "may improve the strategic environment of relatively weak states."[38] This is particularly the case when a powerful, nuclear state adopts a bellicose posture in relation to weaker states. Of the three countries named by President George W. Bush as constituting an "Axis of Evil," only nuclear North Korea was able to escape the full brunt of America's wrath; Iraq was invaded in 2003, and Iran was constantly threatened with military action during President Bush's tenure in office.[39]

---

works as Indicators on Nuclear Weapons Intentions," *The Nonproliferation Review* (Fall–Winter 2000), p. 76.

[35] Ibid., p. 77.

[36] Kroenig, "Exporting the Bomb," p. 114.

[37] Ibid.

[38] Ibid., p. 128.

[39] At the time, the White House claimed that the United States "may face no greater challenge from a single country than from Iran." See The White House, *The National Security Strategy of the United States of America, March 2006* (Washington,

One scenario that is often presented as a highly attractive, if not fully plausible, alternative to the present nuclear regime is nuclear disarmament on a global scale. On a few occasions, there have been high-profile, if not necessarily vociferous, calls by intellectuals and former policymakers advocating the global dismantling of nuclear weapons.[40] While it is attractive in the abstract, the realistic feasibility of such a scenario is highly debatable. Even if nuclear-weapons states dismantle their existing arsenals, their ability to rebuild and deliver nuclear weapons is unlikely to dissipate. In fact, with the possibility of "hair-trigger (nuclear) mobilization" ever present, regional crises and international tensions are much more likely to quickly reach nuclear escalation. Thomas Schelling lays out an alarming scenario. Former nuclear-weapons states, he maintains,

> would have prepared targets to preempt other nations' nuclear facilities, all in a high-alert status, with practice drills and secure emergency communications. Every crisis would be a nuclear crisis, any war could become a nuclear war. The urge to preempt would dominate; whoever gets the first few weapons will coerce or preempt. It would be a nervous world.[41]

Schilling goes on to maintain that the twin forces of deterrence and abhorrence (of what nuclear weapons can do) have prevented the use of nuclear weapons for nearly seven decades. Despite repeated military conflicts involving nuclear weapons states with non-nuclear weapons states, the world has seen an uninterrupted period of "nuclear quiet." "Except for some 'rogue' threats," he claims, "there is little that could disturb the quiet nuclear relations among the recognized nuclear nations. This nuclear quiet should not be traded away for a world in which a brief race to reacquire nuclear weapons could become every former nuclear state's overriding preoccupation."[42]

The possibilities of proliferation are accentuated by the preponderance of international conflicts and tensions across the region. More specifically, a "'higher-than-normal' probability of war in a protracted conflict region encourages states to search for a deterrent capability that ultimately leads to

---

DC: US Government Printing Office, 2006), p. 20. The document went on to warn that if diplomacy failed, American military action against Iran would be unavoidable.

[40] Thomas C. Schelling, "A World Without Nuclear Weapons?", *Daedalus*, Vol. 138, No. 4 (Fall 2009), p. 124.

[41] Ibid., p. 127.

[42] Ibid., p. 129.

nuclear proliferation."[43] This proliferation is often opaque and undeclared, and, by the same token, so are the deterrence and war-avoidance strategies by the states involved.[44]

Insofar as proliferation possibilities in the Middle East are concerned, five broad factors are likely to compel Middle Eastern states to seek nuclear weapons. They include deterring adversaries, compensating for conventional weapons shortcomings, enhancing war-fighting capabilities, garnering domestic political power, and enhancing international standing and leverage, especially in relation to the United States.[45]

Further reinforcing proliferation possibilities are two regional but very different nuclear shadows that loom large across the Middle East, those of Israel and Iran. In addition to environmental and economic considerations, the surge in interest in nuclear energy in the GCC states is due largely to political considerations revolving around Iran and, to a lesser extent, Israel.[46] In fact, as discussed in a number of the chapters here, officials across the region, especially in the Persian Gulf monarchies, have often voiced outright concern and fear over Iran's nuclear program.[47]

Both the Iranian and Israeli cases are subject to extensive treatment in this volume. Iran, for now at least, does not have nuclear weapons, and its leaders, who have at times shown tremendous discord amongst one another, have been uniquely united in denying any intention to acquire nuclear weapons. Iran's rationale for promoting nuclear energy is based on its conception of its "inalienable rights" and its national security.[48] For Iran, the US demand that it abandon nuclear enrichment is "a form of neo-imperialism" designed to preserve Israel's nuclear hegemony in the Middle East and to ensure Western

[43] Saira Khan, *Nuclear Proliferation Dynamics in Protracted Conflict Regions: A Comparative Study of South Asia and the Middle East* (Burlington, VT: Ashgate, 2002), p. 3.

[44] Ibid., p. 59.

[45] Richard L. Russell, "Off and Running: The Middle East Nuclear Arms Race," *Joint Forces Quarterly*, Vol. 58, No. 3 (2010), p. 94.

[46] Susan De Muth, "The Nuclear Debate," *The Middle East*, No. 412 (June 2010), p. 22.

[47] "Where is Iran going to use these weapons?" the Saudi Foreign Minister Saud al-Faisal is reported to have asked rhetorically in a speech in 2006. "If their intention is to bomb Israel, they will kill Palestinians, Syrians, Jordanians, and Saudis as well." Quoted in Sammy Salame and Heidi Weber, "Arab Nuclear Envy," *Bulletin of the Atomic Scientists* (Chicago, IL: September–October 2007), p. 46.

[48] Abbas Maleki, "Iran's Nuclear File: Recommendations for the Future," *Daedalus*, Vol. 139, No. 1 (Winter 2010), p. 107.

dominance over the region.[49] Iranian leaders recognize that nuclear weapons guarantee neither domestic stability nor external security, and that the development or possession or nuclear weapons actually undermine Iranian security. "Even the perception that Iran is pursuing nuclear weapons negatively impacts Iran's power by decreasing its regional influence and increasing its global vulnerabilities."[50]

For its part, Israel's possession of nuclear weapons is "the worst-kept secret."[51] By all accounts, Israel either already had nuclear weapons, or had all the components for a quick assembly, by the start of the June 1967 War.[52] From the very beginning, Israel's nuclear policy has been characterized by opacity, a product of collusion on both sides of the Atlantic. During the Johnson administration, the United States initially insisted on Israel signing the Non-Proliferation Treaty (NPT) and sought to allay Israeli concerns through becoming an open and direct supplier of American weaponry to the Jewish state. But President Johnson gradually dropped his insistence on the NPT and was then replaced in the White House by "an even better friend of Israel."[53] The Nixon White House pursued a similarly ambivalent policy toward Israel and the bomb, ultimately settling on tacit acceptance of Israel's possession of nuclear weapons. It did so through feigning ignorance of Israeli nuclear weapons so long as Israel agreed not to fully assemble a nuclear device—to "develop" a bomb—or to "introduce" one to the Middle East.[54] Avner Cohen in his chapter traces the evolution of this opacity, the utility of which is not lost on other potential nuclear aspirants. This is indeed the direction that Iran might well be moving toward, and it has already been suggested as an attractive option for Saudi Arabia *if* it were ever to make serious moves toward nuclear acquisition.[55]

Questions revolving around proliferation issues are, no doubt, critically important both globally and to the Middle East region, especially in light of

[49] Jeremy Salt, "Israel's Nuclear Weapons: The White House Factor," *Middle East Policy*, Vol. 17, No. 3 (Fall 2010), p. 24.

[50] Maleki, "Iran's Nuclear File," p. 107.

[51] Avner Cohen, *The Worst-Kept Secret: Israel's Bargain with the Bomb* (New York: Columbia University Press, 2010).

[52] Salt, "Israel's Nuclear Weapons," p. 23.

[53] Ibid., p. 27.

[54] Ibid., p. 29.

[55] Richard Russell, "A Saudi Nuclear Option?" *Survival*, Vol. 43, No. 2 (Summer 2001), p. 75.

Western and regional suspicions over Iran's nuclear ambitions. But proliferation is only one piece of a more complex puzzle involving nuclear energy. Much of the recent nuclear renaissance, at least until the 2011 nuclear crisis in Japan, has been inspired by a rediscovery of the virtues of nuclear energy as a safe and efficient alternative to fossil fuel sources.

The case for nuclear energy

For many national policymakers, the appeal of adopting nuclear energy as a viable alternative to existing energy sources rests on two factors. First, nuclear energy is relatively "clean," with nuclear-generated electricity contributing little to greenhouse gas emissions, and can allay concerns about global warming. There is a widespread perception that nuclear energy represents "the most developed and economic low-carbon electricity alternative."[56] Second, nuclear energy can contribute to energy security and reduce or altogether eliminate reliance on fossil fuels for electricity generation.[57] Of course, there is also the question of national prestige, which often weighs heavily in the calculations of national policymakers as they decide to embark on what often amount to showcase projects.

Across the Middle East and the GCC, as the chapters on Saudi Arabia and the UAE document, electricity consumption continues to increase at rates with which existing national infrastructures can hardly keep up. Prompted by forces such as dramatic demographic and population growth, rapid industrialization, and heavily subsidized prices, in the GCC alone the investment required to keep up with increasing electricity demands is estimated at US$7 billion a year for the foreseeable future, so that current levels of installed capacity could be raised by some 80 percent.[58] While China leads the world in average annual per capita increase in electricity consumption, it is followed in the second place by Iran, then by Syria, Kuwait, and Egypt.[59] The rest of the

[56] Harold A. Feiveson, "A Skeptic's View of Nuclear Energy," *Daedalus*, Vol. 138, No. 4 (Fall 2009), p. 66.

[57] Jose Goldemberg, "Nuclear Energy in Developing Countries," *Daedalus*, Vol. 138, No. 4 (Fall 2009), p. 71.

[58] Steffan Hertog and Giacomo Luciani, "Energy and Sustainability Policies in the GCC" (Kuwait: Programme in Development, Governance and Globalisation in the Gulf States, Center for the Study of Global Governance, 2009), p. 8.

[59] Ibid., p. 4.

Middle East tends to have similarly high growth rates as well (see Table 1.1). Qatar and then Kuwait have the highest rates of per capita electricity consumption in the world, followed by the United States, and then the UAE and Bahrain (table 1.2).[60]

Table 1.1: Average annual % increase in per capita electricity consumption

China	13.0
Iran	7.6
Syria	6.3
Kuwait	6.1
Egypt	4.8
Oman	4.3
Bahrain	4.2
India	4.2
Saudi Arabia	4.1

Source: Steffan Hertog and Giacomo Luciani. "Energy and Sustainability Policies in the GCC" (Kuwait Programme in Development, Governance and Globalisation in the Gulf States: Center for the Study of Global Governance, 2009), p. 4.

Table 1.2: Highest per capita electricity consumers in the world (kilowatt hour per person)

Qatar	15,000
Kuwait	14,000
United States	13,000
UAE	11,500
Bahrain	10,500
Japan	7,800
France	7,500
Germany	7,000
UK	6,100
Saudi Arabia	6,100

Source: Hertog and Luciani. "Energy and Sustainability Policies in the GCC," p. 4.

The current nuclear renaissance has given rise to some concerns, many of which are similar to those that initially surfaced in the aftermath of the "atoms for peace" program launched by President Eisenhower in 1953.[61] To begin

[60] Ibid.

[61] One alarmist assessment sees five potential, interrelated risks inherent in the expan-

with, questions about how to handle radioactive waste remain largely unresolved. Vast quantities of enriched uranium will become necessary to run the increasing number of nuclear power plants being planned; the plutonium that is consequently produced in the reactors can potentially be used for weapons; and large amounts of radioactive products will have to be disposed of.[62] The initial promise that future generations will not have to deal with the nuclear waste that is currently being produced has yet to be realized.[63] Technically, solutions such as dry cask storage or geological storage in a repository appear both feasible and safe.[64] But, as the Iranian case demonstrates, storage by itself does not alleviate international concerns about the potential for proliferation. Even the transportation of fuel rods shipped from producing countries and the return of spent fuel (unless they are reprocessed locally), one of a series of options considered in the Iranian case, is fraught with political as well as technical complications.[65]

Similarly, nuclear energy programs move forward only when heavily subsidized by governments, as they are in China and India. As viable commercial ventures, therefore, they remain uncompetitive and produce relatively little energy in comparison to their scale and the level of financial commitment they require on the part of the state.[66] Nevertheless, across the Middle East, where despite economic liberalization initiatives the state continues to remain in charge of key industries, heavy infrastructural investments in electricity-generation plants—as well as hydrocarbons and other strategic industries—are considered the norm. Moreover, as discussed in Mari Luomi's chapter, although the UAE model is often touted as "the gold standard" by senior Western policymakers, including the American Vice President Joseph Biden,[67]

---

sion and spread of civilian nuclear power: safety, sabotage, terrorist theft or the purchase of a weapons or nuclear materials, nuclear weapons proliferation, and the destruction of nuclear facility in a conventional war. Steven E. Miller and Scott D. Sagan, "Alternative Nuclear Futures," *Daedalus*, Vol. 139, No. 1 (Winter 2010), p. 127.

62 Goldemberg, "Nuclear Energy in Developing Countries," p. 77.

63 Robert H. Socolow and Alexander Glaser, "Balancing Risks: Nuclear Energy and Climate Change," *Daedalus*, Vol. 138, No. 4 (Fall 2009), p. 36.

64 Feiveson, "A Skeptic's View of Nuclear Energy," p. 65.

65 Goldemberg, "Nuclear Energy in Developing Countries," p. 78.

66 Feiveson, "A Skeptic's View of Nuclear Energy," p. 60.

67 Pamela Ann Smith, "Nuclear Energy Gains Ground in Arab States," *The Middle East* (June 2009), p. 41.

such deals tend to perpetuate continued external dependence on outside suppliers and monitors.[68] Not surprisingly, so far a number of countries both in the Middle East and beyond—Vietnam, Jordan, Egypt, and the rest of the GCC—have rejected the UAE nuclear deal as a model to emulate.

Regardless of the international regimes to which they may agree in the process, the adoption of "good governance mechanisms" are often seen as essential for those countries interested in pursuing nuclear energy programs.[69] Such mechanisms often include transparency—of varying degrees, of course—and bilateral and multilateral cooperative arrangements with international agencies and suppliers. For an electricity-starved region like the Middle East, the spread of nuclear power could also potentially serve as a point of common interest between former or current foes through collaboration in energy-generating technologies and other similar "peace projects."[70] Such future possibilities could include energy-sharing projects between Iran and its southern neighbors, particularly Kuwait and Bahrain, and even between Israel and Lebanon and Syria.

This volume

The chapters in this volume highlight several paradoxes in the nuclear question in the Middle East. Perhaps of all the regions in the world today, with the possible exception of Africa, the Middle East has been the scene of some of the most profound domestic political changes and international tensions the world has ever seen. The fateful events of 2011 brought tectonic changes to the politics of the Middle East. "Jasmine revolutions" toppled long-serving authoritarian figures in Tunisia and Egypt; civil wars broke out between intransigent government loyalists and opposition forces in Libya; a sectarian-flavored rebellion rocked the Bahraini political establishment to its core; Ali Abdullah Saleh, Yemen's strongman, was brought to his knees after thirty-three years in power; the Syrian state teetered perilously close to the abyss; and, anti-government protests with varying degrees of violence broke out in Algeria, Jordan, Morocco, Saudi Arabia and Oman. As the political landscape was being redrawn in the Middle East, nuclear calamities of unprecedented proportions began to wreak havoc in Japan. Combined, these developments

[68] Goldemberg, "Nuclear Energy in Developing Countries," p. 78.

[69] Miller and Sagan, "Nuclear Power without Nuclear Proliferation?", p. 9.

[70] Thomas Lorenz and Joanna Kidd, "Israel and Multilateral Nuclear Approaches in the Middle East," *Arms Control Today* (October 2010), p. 25.

only added urgency to the need to address some of the nuclear paradoxes in the Middle East.

One of the most fundamental questions that begs further analysis and study has to do with the overall nature and foundations of the political systems that exist in the Middle East. For many decades, the Middle East has remained one of the last bastions of authoritarian rule in the face of successive waves of democratization.[71] Nevertheless, as the events of 2011 made starkly clear, scholarly assumptions about political stability and longevity in Tunisia, Egypt and elsewhere turned out to be far off the mark.[72] So it has often been remarkably difficult to decipher the causes of and the dynamics behind internal decision-making processes, and the primary motivating factors that have prompted political actors to behave one way as opposed to another. Using different theoretical perspectives and with focus on different internal dynamics, Etel Solingen and Maria Rost Rublee explore these domestic sources of nuclear decision making in the Middle East in their respective chapters.

A second paradox has to do with the uneven treatment of different actors in the Middle East based on diplomatic and strategic considerations. The United States and Europe continue to press for progressively more comprehensive sanctions against Iran, despite an absence of proof so far that the Islamic Republic has a nuclear weapons program, while they actively assisted Israel in its acquisition of nuclear weapons decades ago and pretend today that Israel does not have nuclear weapons. At the same time, under the Bush and Obama administrations, the United States entered into a nuclear partnership with non-NPT signatory India that, according to one US Congressional report, does not in any way "restrain India's nuclear weapons program."[73] Do deviations by the nuclear weapons states themselves from some of the central premises of the Non-Proliferation Treaty, especially insofar as disarmament and non-transfer of weapons technology are concerned, call for a revision of the NPT regime? Three chapters in this volume—on Israel, Iran, and Tur-

[71] See, for example, Jason Brownlee, *Authoritarianism in the Age of Democratization* (Cambridge University Press, 2007); and Marsha Pripstein Posusney and Michele Penner Angrist (eds), *Authoritarianism in the Middle East: Regimes and Resistance* (Boulder, CO: Lynne Rienner, 2005).

[72] Stephen King, *The New Authoritarianism in the Middle East and North Africa*, (Bloomington, IN: Indiana University Press, 2009).

[73] Sharon Squassoni, *U.S. Nuclear Cooperation With India: Issues for Congress* (Washington, DC: Congressional Research Service, 2005), p. 3.

key—examine cases in the region with different nuclear trajectories, each of which in its separate way poses challenges to the Non-Proliferation Treaty.

The cases of Israel, Iran and Turkey typify a third and final question: the future trajectory of nuclear developments in the Middle East. Questions involving the possibility of further internationalization of the nuclear fuel cycle, the broader international politics of the region and the globe, both at the macro and meso levels, and renewed concerns over serious environmental risks all present unknown variables for the future direction of nuclear issues in the Middle East. What do recent moves toward the internationalization of the fuel cycle—successfully in the UAE case but not successfully in the case of Iran and the agreement it struck with Brazil and Turkey in May 2010—mean for the Middle East and perhaps for other developing regions? Does the emergence of new actors in regional and global politics—such countries as Turkey, Brazil, India, and even the UAE, Saudi Arabia, and Iran—require adjustments to existing conceptions about and arrangements underwriting international politics, and the role that nuclear developments play in such arrangements? Equally consequential is the 2011 nuclear disaster in Japan, the full consequences of which for the ongoing "nuclear renaissance" across the globe and especially in the Middle East are yet to unfold. In the long run, what does the Japanese nuclear crisis mean for the future direction of the nuclear renaissance, and, more immediately, what lessons can be drawn from it for earthquake-prone countries such as Iran and Turkey? Finally, how viable is the idea of a nuclear weapons free zone in the Middle East?

This volume begins with Etel Solingen's discussion of the domestic sources of nuclear behavior in the Middle East, focusing on the two cases of Iran and Egypt. Solingen draws connections between the general position of a country in the larger international system and the likelihood of its nuclearization. Driven largely by strategies of political survival, she maintains, more inward-looking states, such as Iran, tend to be more prone toward nuclearization, whereas internationalizing states, like Mubarak's Egypt, face greater constraints in doing so. Solingen places the differing nuclear trajectories of Iran and Egypt within the context of ideological and diplomatic rivalries between the two states. What direction the post-Mubarak Egyptian state chooses to follow remains to be seen in the years to come.

Rublee turns her attention to analysis at the individual level with particular reference to the cases of Egypt and Libya, both states in the midst of profound and destabilizing leadership transitions at the time of the book's publication. Rublee employs a social-psychological framework for understanding elite orientation toward nuclear proliferation. She traces the actions and announcements of leaders in both states, arguing that in both cases the decision not to pursue nuclear proliferation over nuclearization ultimately boiled down to choices made by state actors.

In the three chapters that follow we explore the role of nuclear energy across the Persian Gulf. Focusing on the GCC economies, Giacomo Luciani makes a compelling case for the need for nuclear energy as an essential component for the breakneck infrastructural and economic growth taking place across the region. Fossil sources, he maintains, are no longer sufficient to meet the GCC states' growing energy demands. One GCC state that is steadily moving toward complementing its growing energy needs is Saudi Arabia, whose case is examined by Thomas Lippman. The UAE—more specifically the emirate of Abu Dhabi—has already established a fast-track civilian nuclear program. Mari Luomi explores the "defining role" played by France and the US in Abu Dhabi's adoption of nuclear program, as well as the geopolitical ambitions and energy security considerations of the UAE leadership.

Mustafa Kibaroglu, Avner Cohen and Kayhan Barzegar examine the cases of Turkey, Israel, and Iran respectively. Kibaroglu chronicles Turkey's largely unsuccessful efforts at securing access to nuclear technology in order to meet its fast-growing energy needs. Despite its close diplomatic and military alliance with the West, Turkey's nuclear aspirations have been hampered by chronic suspicions that access to nuclear technology could result in its diversion toward military purposes. Similar concerns initially marked Western approaches toward Israel's nuclear (weapons) program. To allay the concerns of allies, and also possible reprisals by adversaries, Israeli leaders adopted a policy of deliberate opacity, discussed in Cohen's chapter. Barzegar, meanwhile, examines the troubled evolution of Iran's nuclear program from its inception to the present, highlighting the employment of nuclear hedging by Iranian leaders in their difficult relationships with the European Union, the United States and Israel.

The book ends with an exploration by Gawdat Bahgat of the possibility of establishing a nuclear weapons free zone (NWFZ) in the Middle East. Drawing lessons from the five existing NWFZs around the world—in Latin America and the Caribbean, the South Pacific, Africa, and Central Asia—Bahgat links the prospects of a NWFZ in the Middle East to the establishment of broader peace and confidence building in the region; enhanced trade and movement of goods and people among the region's fractious, if not warring, actors; and positive security assurances by outside powers. For now, the prospects of a NWFZ in the Middle East are quite remote.

As the chapters in the volume collectively demonstrate, the adoption of a nuclear program for either energy or military purposes is by no means monocausal and is invariably the product of several different factors and considerations. The overall disposition of state leaders regarding their country's level and nature of engagement with the international community is an important predictor of the likelihood of the country's nuclearization, with inward-looking states more likely to embark on nuclearization and internationalizers less likely to do so. In addition to structural factors, decisions made by individual state leaders in response to domestic energy demands, international and domestic threat perceptions, questions of national prestige, and hedging are all critically important in whether or not, and in which direction, a country's nuclear aspirations move. Equally important, as is evident in the Iranian and to a lesser extent the Emirati and Israeli cases, are foreign powers, especially the nuclear weapons states.

In 2011 the Middle East witnessed domestic upheavals of unprecedented proportions. It is unclear whether or not what came to be known as the Arab Spring will have lasting and meaningful consequences for the domestic and international politics of the region. As the initial euphoria of perceived democratization movements across North Africa and the Middle East gave way to the sober realization of brutal authoritarian persistence—as made tragically clear in Libya, Syria, and Bahrain—the imperative of regime survival as an overriding focus of domestic and foreign policy pursuits was once again driven home. Whatever direction Middle Eastern leaders decide to follow on the nuclear question will be similarly based on regime survival calculations. For the time being, the initial indications do not signal radical shifts across the region: in order not to be left behind, and in response to their own needs, the other GCC states are likely to adopt paths similar to that of the UAE; Israel is unlikely to abandon its carefully guarded policy of opacity any time in the near future; Turkey's nuclear aspirations will most likely continue to be viewed with skepticism by its Western allies; Iranian leaders are likely to continue with their often dangerous game of nuclear hedging; and regional circumstances and domestic policy preferences will continue to undermine conditions for a nuclear weapons free zone. It is unlikely, in other words, that questions and tensions around nuclear issues in the Middle East will diminish any time soon.

2

Domestic Sources of Nuclear Behavior in the Middle East

Etel Solingen

Increased concern with the possibility of proliferation waves or epidemics[1] requires a better understanding of what exactly drives states to develop, acquire or renounce nuclear weapons. Various logics can explain the range of decisions associated with these different outcomes.[2] As an early study by Meyer suggested, it is quite likely that some assumptions from different logics are valid;[3]

I would like to acknowledge the help of Mehran Kamrava, participants at the conferences hosted by the Georgetown Center for International and Regional Studies in Doha, Qatar, and Celia Reynolds and Wilfred Wan for research assistance.

1 William C. Potter and Gaukhar Mukhatzhanova, "Divining Nuclear Intentions: A Review Essay," *International Security*, Vol. 33, No. 1 (Summer 2008), pp. 139–69.

2 Etel Solingen, "The Political Economy of Nuclear Restraint," *International Security*, Vol. 19, No. 2 (Fall 1994): pp. 126–69; Etel Solingen, "The Domestic Sources of International Regimes: The Evolution of Nuclear Ambiguity in the Middle East," *International Studies Quarterly*, Vol. 38, No. 4 (June 1994): pp. 305–37; Etel Solingen, *Nuclear Logics: Contrasting Paths in East Asia and the Middle East* (Princeton, NJ: Princeton University Press, 2007).

3 Stephen M. Meyer, *The Dynamics of Nuclear Proliferation* (University of Chicago Press, 1984).

the task is identifying when and why different logics apply. This chapter focuses on domestic sources of nuclear behavior for three reasons. First, the editor's guidelines invited such a focus in an effort to untangle "cause and effects" relations explaining nuclear decisions. Second, as Meyer's early study suggested, all motives of nuclear behavior are, in the end, filtered through the domestic politics within which decisions are made. It is somewhat baffling, therefore, that a *systematic* understanding of domestic effects eluded most work in the area of nuclear proliferation in the last three decades, until fairly recently.[4] Country studies of nuclear aspirants often provided *ad hoc* depictions of the domestic scene but these were often drawn without reference to an overarching comparative framework. Cleavages between "moderates" and "hardliners" were often identified on the basis of inductive "who's who" analyses, lacking an underlying logic for what makes leaders, institutions, ruling coalitions (or their opponents) "moderates" or "hardliners." Third, I build on previous work to draw attention to some recurring regularities in the analysis of domestic sources of nuclear decisions. These regularities relate to ideal-typical models of political survival that leaders embrace to gain or retain power. As Katzenstein and Sil argue, this has been the most under-theorized dimension in explaining nuclear choices until recently.[5]

Different models of domestic political survival provide important information regarding nuclear decisions in the post-1960s era, after the inception of the Non-Proliferation Treaty (NPT). In particular, these models entailed different orientations to the global political economy and its associated economic, political, and security institutions, and had different implications for nuclear choices in various regions of the world. On the one hand, leaders advocating economic growth through integration in the global economy ("internationalizing" models) had incentives to avoid the costs of nuclearization. Leaders vary in their tolerance for domestic and international, political and economic (including opportunity) costs entailed by nuclear weapons. What specific aspects of models emphasizing economic growth and openness to the global economy as tools of political survival make leaders more receptive to denuclearization than others? The need to appeal to foreign investors with an interest in domestic economic growth and stability; the related need

[4] For an early statement regarding this neglect, see Solingen, "Political Economy of Nuclear Restraint."

[5] See their assessment of Solingen, *Nuclear Logics*, in Peter J. Katzenstein and Rudra Sil (eds), *Beyond Paradigms: Analytic Eclecticism in the Study of World Politics* (New York: Palgrave, 2010), p. 75.

to reassure neighbors in order to preserve regional cooperation, stability, and attractiveness to international economic actors; securing access to international markets for exports, capital, technology, and raw materials; the related aversion to risking reputational losses at home and abroad for uncertain nuclear gains; and the costs of alienating domestic agents of internationalization—both within and outside state structures—which would be adversely affected by nuclear weapons development.

Clearly, there are several causal pathways linking renunciation of nuclear weapons to models emphasizing economic growth through global integration. Nuclearization has negative implications for enhancing exports, economic competitiveness, macroeconomic and political stability, and global access—all objectives of internationalizing models. Nuclearization can also strengthen state bureaucracies and industrial complexes—particularly military ones—opposed to economic transformation. For all these reasons, internationalizing leaders, who might otherwise have considered nuclear weapons, chose to avoid them, often as part of a broader program of internationalization designed to strengthen market-oriented forces, leaders, and institutions—state and private—favoring export-led growth and industrialization.

On the other hand, leaders relying on inward-looking models had greater tolerance—and in some cases strong incentives—for development of nuclear weapons. Nuclearization entailed fewer costs for inward-looking leaders and coalitions whose political platforms were rooted in mistrust for international markets, investment, technology, and institutions. Such coalitions protected uncompetitive national industries, sprawling state enterprises and ancillary military-industrial and nuclear complexes. Nuclearization entailed considerable domestic advantages for opponents of internationalizing models in inward-looking, import-substituting regimes favoring extreme nationalism, religious radicalism, or autarky. Such leaders and their political allies often relied on extreme language to compel and threaten regional adversaries, wielding potential nuclear and other Weapons of Mass Destruction (WMDs) as means to coerce and intimidate. Statements such as North Korea's repeated threats to turn Seoul and Tokyo into a "sea of fire," Saddam Hussein's threats to incinerate Iran, Saudi Arabia and Israel, and similar Iranian threats to "wipe out" Israel have been much rarer in domestic political contexts driven by internationalizing objectives. Most inward-looking nuclear aspirants have been NPT members who misled the International Atomic Energy Agency (IAEA) and violated NPT commitments. In particular, they were generally more promiscuous regarding state-directed or state-endorsed global dissemination of sensitive nuclear technologies, fueled largely by the very structure of the domestic models of political survival that sustained them in power. Militarily sensitive exports, as with North Korea, have been a source of income for such models, unable to otherwise accumulate resources from competitive civilian exports to the rest of the world.

Thus, whereas inward-looking models might have regarded nuclear weapons programs as assets in the arsenal of building regime legitimacy, internationalizing models built legitimacy on successful and competitive economies capable of distributing public goods to broad sectors of the population. This contrast is most evident when comparing the Middle East with East Asia. As a region, the Middle East gravitated toward the inward-looking end of the spectrum for decades, accounting for most cases of nuclearization since the 1960s. Many Middle East leaders relied on models of self-sufficiency and nationalism for their political survival, and had stronger domestic incentives to seek nuclear weapons. Egypt, a leader of the Arab world, might be considered an anomaly, avoiding nuclearization after President Sadat launched *infitah* (opening to the global economy), a political-economy strategy that was consistently challenged at home and often flawed in its execution, and hence had uneven results.

In contrast to the Middle East, nuclearization has been much less politically attractive and far more costly for most East Asian leaders since the 1970s, except for North Korea, the autarky-seeking anomaly. Heavy regional concentration of internationalizing models in East Asia—economic growth via integration in the global economy—reinforced each state's incentives to avoid nuclearization. This strategy's mirror image—heavy regional concentration of hyper-nationalist, economically protectionist and militarized models throughout the Middle East—exacerbated incentives to develop nuclear weapons. In other words, regions more highly integrated in the global economy enhanced the prospects of internationalizing models that restrain nuclearization, whereas regions less integrated into the global economy posed more serious difficulties for denuclearization. As the preceding statement makes clear, a causal logic that enhances the prospects for nuclearization does not imply a deterministic outcome but rather a proclivity or tendency. A case that disconfirms the general tendency is not only empirically possible but even desirable, as it dispels concerns with tautology.

The identification of core models of political survival underlying the domestic politics of nuclear aspirants provides a systematic tool for evaluating such tendencies. Both qualitative and quantitative studies across many regions validate the need to pay greater attention to the links between political-economy models and nuclear choices. Some quantitative studies,[6] for instance, found support for the propositions that

1. For all countries that explored the nuclear option, trade liberalization provides the most statistically significant predictor of whether or not they acquire nuclear weapons;
2. The process of economic liberalization is associated with a reduced likelihood of exploring nuclear weapons;
3. Economic openness "has a statistically significant negative effect" on exploring, pursuing, or acquiring nuclear weapons;
4. Economic liberalization has a positive and statistically significant effect on advancing a nuclear weapons free zone (NWFZ) treaty ratification.

The connection between models of political survival and nuclear policies since the 1960s also finds support in systematic qualitative studies, across different regional security contexts, diverse associations with hegemonic powers, and successive leaderships within the same state. First, every case of denuclearization entailed a domestic evolution toward internationalization. Of all nuclear aspirants in the last three decades, not one endorsed denuclearization—fully and effectively—under domestic regimes that shunned integration in the global political economy. Only leaders and ruling coalitions advancing their political survival through export-led industrialization undertook effective commitments to denuclearize (Japan, Taiwan, South Korea, Egypt under Sadat, South Africa, Brazil and Argentina, Algeria and Libya in 2003). Nuclear decisions were invariably concomitant with a broader shift toward internationalization in economics and security, even if the shift was not equally sustained in all cases.

Secondly, where internationalizing leaders and coalitions became stronger politically, as in Japan, South Korea, and Taiwan, the departure from nuclear claims was maintained even as their security context deteriorated (as in the Korean peninsula and the Taiwan Straits, at various points). The relationship between politically stronger internationalizers and the timing of rolling back nuclear ambitions was also evident in Argentina under Menem, Brazil under Fernando H. Cardoso, Spain (with accession to the NPT preceding EU membership) and South Africa, among others.

6 Sonali Singh and Christopher R. Way, "The Correlates of Nuclear Proliferation: A Quantitative Test," *Journal of Conflict Resolution*, Vol. 48, No. 6 (2004): pp. 859–85; Matthew Fuhrmann and Xiaojun Li, "Legalizing Nuclear Restraint: When Do States Enter Nuclear Weapons Free Zones?" (Managing the Atom Working Paper Series, Harvard University, 2008).

Thirdly, where leaders and coalitions favoring internationalization were weaker, as was the case historically in Argentina and Brazil until the early 1990s, in Iran and other places, they were more politically constrained in curbing nuclear programs.

Fourthly, most defiant nuclear courses (albeit not all) have been unmistakably embraced by autarkic or inward-oriented models, from Perón in Argentina to Vargas in Brazil, Nasser in Egypt, and leaders in North Korea, Iraq, Iran, Syria, possibly Algeria and pre-2003 Libya. Indonesia's Sukarno—suspected of coveting nuclear weapons—fits this model as well, as do reported but unconfirmed designs by the Burmese junta and Venezuela's Hugo Chavez.

Fifthly, even advocates of internationalizing models may have to contend with dangerous regions where neighboring leaders endorse alternative economic and nuclear policies. This problem is less severe in East Asia, where export-led industrialization spread from Japan to most neighboring countries (except North Korea), than in the Middle East. In other words, whether the regional center of gravity is internationalizing or inward-looking matters for the individual calculations of states within that region.

Logics in the plural

As gleaned from the experience of an overwhelming number of cases, domestic political-economy models of political survival should be more fundamental considerations in explaining nuclear choices than they have been hitherto. These are not merely afterthoughts or residual factors. Their omission has important implications. A "missing" or "omitted variable" may lead to an overestimation of other causal variables, granting them too large an effect on the outcome while rendering at least some of their effects spurious. Without taking into account political survival models one may not properly understand nuclear behavior or estimate the actual effects of other variables, such as balance of power, international norms and institutions, or democracy. This point bears repetition and elaboration, as it warns against superficial interpretations of the argument and cautions against simplistic understandings of "cause and effects" relationships.

First, introducing a previously omitted variable does not imply that other variables are rendered irrelevant, only that we are better able to understand their relative impact on nuclear choices, as *Nuclear Logics* made explicit.[7]

[7] Solingen. *Nuclear Logics.*

Domestic models of political survival may not explain all cases well but they consistently shed light on how domestic orientations to the global political-economy create different constraints, incentives, receptivity, and compliance patterns, and condition the role that international power, institutions and democracy have played in nuclear decisions. Such models thus help explain why security dilemmas are sometimes seen as more (or less) intractable, why some states rank alliances higher than self-reliance but not others, why nuclear weapons programs surfaced where there was arguably little need for them (Libya, the Southern Cone and South Africa), why such programs were obviated where one might have expected them (Vietnam, Singapore, Jordan and many others) and how leaders filter different concepts of security through their preferred approach to political survival.

Second, Katzenstein and Sil are correct that *Nuclear Logics* was eager to give domestic models of political survival more of a hearing than they had received; that this focus did not imply an exclusion of other causal arguments; and that indeed the latter were used where necessary to arrive at a satisfactory answer to the question at hand.[8] They note, in particular, the explicit recognition of "Logics in the Plural" in the book's very title, but also the need to avoid overestimation of theories such as neorealism, which rely on self-help and international anarchy as the core organizing principle of international relations.

Third, domestic models of political survival can be considered *filters* through which other considerations operate, tilting the balance of incentives and (dis)incentives to acquire nuclear weapons in one direction or another. This point is often missed by reductionist interpretations of the argument, which erroneously assume that taking into account domestic models implies lack of attention to external dimensions, including threats. One should consider domestic models as something akin to a lens that refracts, enhances or redirects external stimuli according to the incentives embedded in each model. Properly interpreting these models as filters of a wide-ranging set of domestic, regional, and global opportunities and constraints also requires tolerance for complexity. As Philip Tetlock's masterful treatise on expert political judgment suggests, parsimony can be the enemy of accuracy, a substantial liability in the effort to understand phenomena in international politics, including nuclear behavior.[9]

[8] Katzenstein and Sil (eds), *Beyond Paradigms*, p. 75.

[9] Philip E. Tetlock, *Expert Political Judgment: How Good is it? How Can We Know?* (Princeton, NJ: Princeton University Press,, 2005).

Fourth, evidence for domestic political considerations in nuclear outcomes is much harder to garner. Leaders are far more likely to cast decisions favoring or rejecting nuclear weapons as "reasons of state," invoking national security, international institutional incentives, and normative considerations (for or against such weapons) rather than ulterior domestic political motivations or expediency. Hence, available historical sources document "reasons of state" more frequently and thoroughly, loading the evidentiary dice against domestic political justifications. This methodological difficulty should not, however, preclude proper exploration of important sources of nuclear behavior. It should force us, instead, to redouble our efforts to ensure that we do not simply look under the proverbial lamp for a key lying elsewhere.

Fifth, the general findings reported in the preceding section suggest there is value in identifying more discrete markers, shortcuts, or rules-of-thumb that explain the motivations of leaders, ruling coalitions and their opponents. At the same time, and reiterating an earlier point, the proposition that domestic orientations to the global economy and nuclear policy are linked is probabilistic, bounded, and refutable. It is probabilistic, as are most arguments in the social sciences, because it does not suggest an inevitable or deterministic outcome. And it is bounded in three ways.

First, resistance to the global economy may provide only a near-necessary but not a sufficient condition for the development of nuclear weapons programs.[10] For instance, not all Middle East leaders who rejected the global economy as a favored platform of domestic political survival have also developed nuclear weapons. Ironically, *Nuclear Logics* placed Syria as an anomaly that presumably did not pursue nuclear weapons despite its inward-looking, nationalist, protectionist, and militarized political-economy. The book's presumption that Syria abided by its non-proliferation commitments was based on the IAEA's determination of compliance with NPT and IAEA commitments prior to its publication. Since 2008, however, the IAEA has begun questioning the extent to which Syria's activities have been NPT-compliant. In so doing, it has validated yet another crucial case of an inward-looking Middle East model prone to pursue nuclear weapons.

Secondly, the proposition linking the two basic models with nuclear behavior is also bounded by regional considerations. The extent to which countries in a given region share a congruent orientation toward the global political economy (either positive or negative) modifies domestic preferences on nuclear issues in each country. For instance, in recent decades East Asian—

[10] Few social science theories meet the standards of necessity and sufficiency.

including Southeast Asian—leaders and ruling coalitions generated one of the most internationalized regions in the world. This fact reinforced individual incentives of each East Asian leader to maintain a collective trend away from nuclearization for the sake of common regional stability, stable foreign investment, and domestic growth. North Korea's economic closure made it more impermeable to the positive regional synergies operating on East Asian states. By contrast, an entrenched inward-looking model at the regional level (under Nasser, in particular) in the Middle East burdened individual efforts by other states, which would have otherwise deepened their orientation to the global economy. The disincentives to develop nuclear programs that often operate in more open political economies were thus weaker in this region. Individual leaders or ruling coalitions that might have otherwise favored integration with the global economy—for example Jordan or Lebanon for many years—faced an unwieldy neighborhood that actively discouraged it.

Thirdly, the argument emphasizing competing political-economy models may be bounded by temporal sequences in the acquisition of nuclear weapons. For instance, the incentives of a globally integrated political economy may operate more forcefully on cases where nuclear programs have not yet yielded nuclear weapons, as seems to have been the case in East Asia as well as in Argentina, Brazil, Spain, and others. Such incentives may have less impact once nuclear thresholds—often in the form of nuclear tests—have been crossed. It seems far more politically costly for leaders and ruling coalitions to eliminate *existing* nuclear weapons than it would be to abandon programs that have not yet come to fruition. This would be the case even for internationalizing advocates and particularly for those surrounded by inward-looking regimes, an argument often raised regarding India and Israel.

In sum, the framework discussed thus far provides a roadmap for considering the conditions under which its expectations might be corroborated or refuted. Various theories suffer from indeterminacy, tautological reasoning, or post-hoc rationalization. For instance, neorealism's concepts of self-help and relative power can drive states to a wide array of nuclear choices, from straightforward acquisition to alliances to straightforward renunciation, and a myriad of possibilities in-between. Its tenets are thus indeterminate, rendering them hardly a reliable guide to *anticipate* what states might actually do. Anomalies in neorealism habitually require additional information unrelated to international power balances. As Betts argues, insecurity is not a sufficient condition for acquiring nuclear weapons; many insecure states have not done so.[11] In

[11] Richard Betts, "Universal Deterrence or Conceptual Collapse? Liberal Pessimism

other words, such theories are afflicted with the problem of multifinality, suggesting many outcomes consistent with a particular value of one variable. It is unclear to this day what the precise underlying measures of relative power that lead to nuclearization are. Yet such clarity is a *sine qua non* to avoid circularity and *ex post facto* rationalizations (such as "state x went nuclear because of acute insecurity," whereby the acuteness threshold is detected by a nuclear test). Arguments must be cast in falsifiable and *a priori* terms, and enable more clearly defined and testable propositions that can disconfirm the theory.

Models of political survival help explain why internationalizing states perceive a regional environment of inward-looking models as more threatening than one populated by like-minded internationalizers. A massive two-volume study by Potter and Mukhatzhanova finds the argument that competing models of political survival shaped positions on nuclear weapons particularly convincing for Egypt, as well as for Saudi Arabia, Turkey, South Korea, Japan, and to a large extent for Iran, South Africa, Taiwan and Ukraine.[12] For many of these cases, the authors maintain, a political-economy focus on outward-oriented versus inward-oriented models is helpful in accounting for much of the variation in nuclear restraint across and within states over time. The evidence provided in twelve dedicated case-study chapters, they argue, offers more support for this theory than is made explicit in the case-study analysis. This is particularly remarkable because proper analyses of domestic politics are especially difficult, chiefly but not only in non-democratic contexts. The standard neorealist argument, by contrast, is far less demanding but is neither necessarily freer of methodological problems nor more accurate, as these volumes confirm.[13]

Iran

The framework described in the preceding section has been applied to explain Iran's road to nuclear defiance.[14] As an important study argued, "the Iranian

---

and Utopian Realism" in Victor A. Utgoff (ed.), *The Coming Crisis: Nuclear Proliferation, U.S. Interests, and World Order* (Cambridge: MIT Press, 2000), pp. 51–86.

12 William C. Potter and Gaukhar Mukhatzhanova (eds), *Forecasting Nuclear Proliferation: The Role of Theory, Volume 1* (Stanford University Press, 2010); Potter and Mukhatzhanova (eds), *Forecasting Nuclear Proliferation in the 21st Century: A Comparative Perspective , Volume 2* (Stanford University Press, 2010), p. 343.

13 For an explanation of different theories' distinct methodological difficulties, see Solingen, *Nuclear Logics*.

14 See Solingen, "Political Economy of Nuclear Restraint"; Solingen, "Domestic Sources of International Regimes"; Solingen, *Nuclear Logics*.

system is one in which ... domestic factional dynamics drive policy debates and policymaking."[15] The 1979 Islamic revolution adopted a quintessentially inward-looking model of political control expropriating industry and applying state control over all major industries including oil, petrochemicals, foreign trade, banking, mining, insurance and transportation; creating large state enterprises, and powerful tax-exempt foundations (*bonyads*) dominated by radical clerics or former military officers; and condemning so-called "Western" international, including economic, institutions. The state came to control over 80 percent of the economy, among the largest state sectors in the world, putatively redistributing wealth from the private to the public sector to attract populist support. The reality departed from these stated objectives, however, as, according to Halliday,[16] power became concentrated in a *nomenklatura* of about 5,000 clerical and lay officials from heavily intermarried families ruling over the state, army and the rentier economy as an extended family. The model privileged its economic allies while largely undermining non-oil private-sector exporters. Ironically, Iran became more dependent on oil and gas exports 30 years after the revolution than it had been prior to it (about 70 percent of the central budget and about 80 percent of foreign exchange is accounted for by revenues from the oil industry).[17] The Swiss KOF Index of Globalization in 2009 found Iran to be among the least globalized economies, near Rwanda and Niger.[18]

The nationalizing, statist, inward-looking strand in Iranian politics was intermittently challenged by a broad coalition of reformist, "economy first," or "pragmatic" advocates of privatization, increased trade with Europe and Asia, a utilitarian over an ideological approach to foreign policy, and decreased tension with international institutions.[19] This strand has been

[15] David E. Thaler *et al.*, *Mullahs, Guards, and Bonyads* (Santa Monica, CA: RAND Corporation, 2010).

[16] Fred Halliday, *The Middle East in International Relations: Power, Politics and Ideology* (New York: Cambridge, 2005), p. 293.

[17] Suzanne Maloney, "Sanctioning Iran," *The Washington Quarterly*, Vol. 33, No. 1 (2010): pp. 131–47; Jahangir Amuzegar, "Iran's Economy in Turmoil," *International Economic Bulletin* (Washington, DC: Carnegie Endowment for International Peace, March 2010).

[18] KOF Swiss Economic Institute, Press Release (Zurich: January 27, 2009), at http://globalization.kof.ethz.ch/.

[19] Roger Owen, *State, Power, and Politics in the Making of the Modern Middle East* (New York: Routledge, 2000), p. 141; Hamid Ahadi, "Government Pushes Largest

besieged by the competing inward-looking populist agenda, particularly since President Mahmoud Ahmadinejad's ascent to power, favoring rentierism and patronage for the benefit of allies, most notably among the Revolutionary Guards. Privatization became a way to transfer state enterprises to protégés in *bonyads* and state banks.[20] Even under President Khatami's economic team, basic commitments to a 1970s-style statist economy were retained.[21] By the 2004 parliamentary elections, characterized by low turnout, reformists had been decimated through imprisonment, torture and popular apathy. Inward-looking policies affected oil production and exacerbated unemployment, inflation, corruption and declining living standards. For all the populist rhetoric, income inequality remained at roughly the same level since 1979, as measured by Gini indices.[22] Meanwhile, Revolutionary Guards enriched their own coffers by importing banned and expensive commodities into private ports under their exclusive control. Closure to the global economy was vital to these groups' prosperity. Accordingly, one of Ahmadinejad's economic advisors, Abumohammed Asgar-Khani, pressed him to "go with techno-nationalism," import-substitution, state control of the economy, and keeping foreign investors out.[23]

Private Bank Out of Business," *Rooz* (October 10, 2006), at http://www.roozonline.com. For an account of three competing ideologies—conservative, reformist and secular—see Mehran Kamrava, *Iran's Intellectual Revolution* (Cambridge University Press, 2008). For an expanded discussion of shifting coalitions in Iranian politics, see also Solingen, *Nuclear Logics*; Thaler *et al.*, *Mullahs, Guards, and Bonyads*; and Shahram Chubin, "The Iranian Nuclear Riddle after June 12," *The Washington Quarterly*, Vol. 33, No. 1 (2010): pp. 163–72.

20 Mehdi Khalaji, "Iranian President Ahmadinezhad's relations with Supreme Leader Khamenei," *Policy Watch* no. 1147 (Washington Institute of Near East Policy, September 12, 2006); Abbas Milani, "U.S. Policy and the Future of Democracy in Iran," *Washington Quarterly*, Vol. 28, No. 3 (2005): pp. 41–56.

21 Meghan L. O'Sullivan, *Shrewd Sanctions: Statecraft and State Sponsors of Terrorism* (Washington, DC: Brookings Institution, 2003).

22 A Tehran analyst reported that according to the Social Security Organization 30 percent of the population was living under the poverty line, whereas the top 20 percent controlled 80 percent of the total wealth and 50 percent of national income. See "Ahmadinejad's Domestic Troubles," The Intelligence Summit (October 4, 2006) at http://intelligence-summit.blogspot.com.

23 Michael Ignatieff, "Iranian lessons," *New York Times Magazine* (July 17, 2005), p. 51; Arms Control Association, "History of Official Proposals on the Iranian

The growing dominance of the inward-looking faction has been accompanied by growing defiance in the nuclear arena.[24] According to Chubin, Iran's nuclear program was clearly intended to buttress the regime's faltering legitimacy and to appeal to the revolutionary ranks.[25] It was a program "in search for a rationale" not rooted in security imperatives but in nationalism, prestige, and domestic drivers. In December 2002 the existence of two previously undeclared Iranian nuclear facilities (Natanz and Arak) was revealed and later confirmed by IAEA Director General Mohamed ElBaradei. The IAEA urged Iran's regime to fully comply with the NPT and sign the Additional Protocol.[26] The international response intensified after June 2003, when the IAEA Board of Governors derestricted a report that outlined how Iran had failed to fulfill its safeguards obligations.

The EU3 (Britain, Germany and France) took center stage following the 2003 IAEA report and resolution requesting that Iran fully declare its activities and allow unrestricted access by October 31, 2003. The EU3 continued to engage Iran over the next few years, heading multilateral efforts to curb Iran's nuclear program. This process did not proceed smoothly, however, as a stream of IAEA reports and resolutions (in March, June and September of 2004) underlined that Iran declined to follow through in its obligations to fully disclose its nuclear history. Furthermore, disagreement on the actual scope of the suspension of uranium enrichment (a technology acquired clandestinely through the Abdul Qadeer Khan network) and reprocessing activities remained. Yet the US agreed to remove its veto on Iran's application to the World Trade Organization as a positive incentive, leading to the 2004 Paris Agreement whereby the EU3 offered valuable technology, fuel, trade conces-

---

Nuclear Issue" (2009) at http://www.armscontrol.org/factsheets/Iran_Nuclear_Proposals.

24 See IAEA documents on Iran at http://www.iaea.org/NewsCenter/Focus/IaeaIran and at http://isis-online.org/isis-reports/detail/iaea-iran-safeguards-report-shut-down-of-enrichment-at-natanz-result-of-stux/html.

25 Shahram Chubin "Does Iran Want Nuclear Weapons?", *Survival*, Vol. 37, No.1 (Spring 1995), pp. 86–104 and *Iran's Nuclear Ambitions* (Washington, DC: Carnegie Endowment for International Peace, 2006).

26 Additional Protocols were enacted by the IAEA in the 1990s in order to "strengthen the effectiveness and improve the efficiency of the safeguards system". See "IAEA Safeguards Overview: Comprehensive Safeguards Agreements and Additional Protocols" at http://www.iaea.org/Publications/Factsheets/English/sg_overview.html.

sions and security help in exchange for Iran's voluntary (and temporary) suspension of its enrichment activities and implementation of the Additional Protocol.

The Khatami administration—elected with broad reformist support—suspended enrichment and accepted the Additional Protocol in principle, in an effort to accommodate international concerns. Yet reformists remained challenged by hardliners, politically and in some cases through death and threats of elimination. Another wave of anti-reformist measures came through manipulated "elections" to the Majles (2004) and presidential elections (2005).[27] Ahmadinejad's ascent to the presidency shaped the next phase in the international response to the nuclear program. After rejecting a EU3 offer (on the grounds that it did not acknowledge Iran's right to enrichment), Iran resumed uranium conversion activities, thus canceling the Paris Agreement. With support from the EU3, Russia offered renewed negotiations for a joint enrichment venture (November 2005), and later, with support from the US and China, Russia offered to operate civilian nuclear facilities in Iran. Iran's rejection of those proposals pushed the IAEA Board to refer Iran's dossier to the UN Security Council (February 2006). China, Russia and the US joined the EU3's efforts in earnest in June 2006, presenting Iran with conditional offers of economic, technological and security incentives to curb its nuclear program. Once again, revealing the entrenched power of the more radical inward-looking faction, Iran rejected proposals calling for a cessation of its enrichment activities.[28]

By contrast, reformers and pragmatists recognized that "integration into the international order and the global economy mandates accepting certain restrictions on [Iran's] nuclear programme."[29] Fifty prominent Iranian econo-

[27] Chubin, "The Domestic Politics of the Nuclear Question in Iran," (Aspen European Strategy Forum (Berlin 2010) at http://www.carnegieendowment.org/static/npp/pdf/chubin-aspen.pdf; Chubin, "The Iranian Nuclear Riddle," pp. 163–72.

[28] It is not possible in the context of this brief overview to elaborate on the diversity *within* the radical and reformist groups, or differences between so-called "pragmatic conservatives" (Larijani, Qalibaf and others) and "principlists" (Ahmadinejad and his IRGC/Basij allies), but see among others Thaler *et al., Mullahs, Guards, and Bonyads*. Some consider the differences between the latter two groups to be mainly personal rather than ideological, whereas the views of reformists are considered to be substantively different from the other two groups. Evolving tension between Khamenei and Ahmadinejad are often traced to a struggle for control rather than to dramatic policy differences.

[29] Ray Takeyh, *Hidden Iran* (New York: Times Books, 2006).

mists wrote an open letter to Ahmadinejad, the first of its kind, emphasizing that:

> the adoption of expansive monetary policies, and the existence of tension in Iran's international relations, [have] caused the investment atmosphere in the country to decline on a daily basis. By increasing risks and the expenses of the private sector, both foreign and domestic, private investment in the country has fallen sharply, while human and material resources are fleeing the country.[30]

They also warned against the dangers of economic sanctions, the consequences of which, they argued, were predictable:

> The rise in investment risk, disorder in foreign trade, reduction in the rate of profit from existing production capacity, slowdown in collecting capital, and the creation of deep inflation (stagflation) are among the harmful [effects] the country will have to bear in the event of a lack of constructive interaction with the global economy... We insist your excellency and the regime's officials attend to a fundamental review of your economic policies and programs for the country.

A group of powerful businessmen, the Islamic Coalition Party, met with a senior member of the Supreme National Security Council to urge moderation in the country's nuclear policies and avoidance of further harm to the economy. Others advocated a "rational" and "transparent" foreign policy favoring "dialogue and diplomacy over adventurism...and chicanery."[31] Domestic criticism of the nuclear program, however, remains forbidden, with the imprisonment of former negotiator Hossein Mousavian serving as a powerful reminder.[32]

Iran failed to respond to the July 2006 Security Council Resolution 1696 demanding the suspension of enrichment and threatening sanctions for noncompliance, leading to the United Nations Security Council (UNSC) application of sanctions in December 2006. UNSC Resolution 1737 targeted both trade-related and financial elements of Iran's nuclear program. Resolutions 1747 (March 2007), 1803 (March 2008) and 1835 (September 2008, which reaffirmed the previous three rounds of sanctions) expanded the list of sanctioned Iranian individuals and entities. Another round of positive incentives by the EU3, and later the P5+1, in 2008—offering to suspend sanctions—were rejected by Iran.

30 "Economists send Ahmadinejad letter urging economic policy change," *E'temad-e Melli* (Tehran: June 21, 2006).

31 Chubin, "The Domestic Politics of the Nuclear question in Iran."

32 Criticism of Ahmadinejad's economic policies has also cost an Economy and Finance Minister and a Central Bank Governor their posts.

Meanwhile the IAEA found evidence that Iran was in violation of its required nuclear reporting commitments and that it had misled inspectors with false information regarding: (1) its centrifuge procurement efforts and activities; (2) shaping uranium metal into hemispheres used in developing explosive cores for nuclear weapons (while refusing to provide the IAEA with a document to that effect); (3) testing conventional high explosives for use in an implosion-type nuclear weapon at Parchin; (4) evidence found in an Iranian laptop about designs to produce uranium tetrafluoride ("Green Salt"); (5) design of a missile re-entry vehicle and payload for Iran's medium-range Shahab-3 ballistic missile; (6) evidence that the uranium project, high-explosives tests, and re-entry vehicle design appear to have "administrative connections" according to documentation found on the laptop; and (7) experiments with separation of polonium-210 suitable for helping to trigger a nuclear chain reaction in certain types of nuclear weapons.[33]

Notwithstanding these IAEA findings, the newly elected administration of Barack Obama, in a significant departure from past US policy, offered to engage in negotiations without the precondition of suspending enrichment. Two personal letters from Obama to Khamenei received no response. Instead, former senior Iranian officials, including a former president, told analyst Karim Sadjapour that in private discussions Khamenei declared, "*Ma doshmani ba Amrika ra lazem dareem*," that is, "We need enmity with the United States." And a month before what many considered a tainted presidential election in June 2009, Khamenei declared that Iran would face a national "disaster" if a candidate who attempted to thaw relations with America came to power.[34]

As the possibility of such a development came about in the context of the 2009 presidential elections, the regime responded with severe domestic repression of protests triggered by what were widely considered illegitimate results awarding Ahmadinejad a majority of the votes.[35] This was followed by revelations of yet another undeclared Iranian enrichment facility (near Qom) in

[33] "Facts Sheet and Briefs," Arms Control Association at http://www.armscontrol.org/factsheets/Iran-IAEA-Issues.

[34] Karim Sadjadpour, "The Sources Of Soviet Iranian Conduct," *Foreign Policy* (Washington, DC: November 2010) at http://www.foreignpolicy.com/article/2010/10/11/the_sources_of_ssoviets_iranian_conduct.

[35] Ian Black, "Resilient Iranians still dream of a new revolution," *The Guardian* (London:December11,2009athttp://www.guardian.co.uk/world/2009/dec/11/iran-new-revolution-dream.

September 2009. Iran's subsequent rejection of a uranium-for-fuel exchange deal toward the end of 2009, and an IAEA report in February 2010 suggesting (for the first time) that Iran may be actively pursuing nuclear weapons capability, has led, once again, to deepening sanctions backed by all permanent members of the UNSC, including China and Russia. A highly respected NGO advances that there is "extensive evidence assembled by the IAEA and Western nations of secret Iranian work on developing the capability to build nuclear weapons."[36] Meanwhile, the nuclear program is said to have assumed even more crucial domestic importance than ever before for a regime tenaciously seeking to survive in power.[37] Various factors are likely to influence Iran's nuclear policy in the future, including the domestic effects and distributional consequences of international sanctions. In the short term, it appears that the nature of the dominant model of political survival embraced by the hardline leadership will shape and define the ultimate contours of Iran's nuclear program. As Kamrava and others suggest, the 2009 elections and their aftermath reflect growing polarization between large segments of Iranian society and those who dominate the state; a progressive militarization of the political system; Khamenei's deep commitment to the dominant faction, and the emergence of new "rules of the game."[38]

Egypt

Tracing Egypt's nuclear abstention in the last four decades to models of political survival seems risky at first sight. Egypt provides in some ways a "most

[36] "Briefing notes from February 2008 IAEA meeting regarding Iran's nuclear program," Institute for Science and International Security (ISIS Report, April 11, 2008) at http://isis-online.org/uploads/isis-reports/documents/IAEA_Briefing_Weaponization.pdf.

[37] Chubin, "The Domestic Politics of the Nuclear Question in Iran."

[38] Ironically, the groups most forcefully proclaiming Iranian "unity" tend to be Iran's regime hardliners on the one hand, and a few external critics of Iran who deny consequential, enduring cleavages in Iranian domestic politics (including on the nuclear issue) on the other. Both groups maintain this claim of unity even in the aftermath of widespread turmoil and repression since June 2009. In academic frames of reference, neorealist and some constructivist frameworks converge on a view of Iran as a unified entity driven either by presumed external imperatives of relative power or symbolic motives and organic identity. For more on this see Mehran Kamrava, "The 2009 Presidential Elections and Iran's Changing Political Landscape," *Orbis*, Vol. 54, No. 2 (2010), pp. 400–12.

likely case" for seeking nuclear weapons under a different—neorealist—logic, where the domestic nature of states does not matter and only "relative power" between/among adversaries does. First, Egypt had deep mistrust of British intentions in the 1950s and was geographically adjacent to a presumed nuclear weapons state, Israel. Second, over the last few decades Egypt also faced other competitors in the regional balance-of-power game such as Iraq, Iran, Algeria and adjacent Libya, as well as Syria, several of which either abstained from joining the NPT for many years or violated their non-proliferation commitments, triggering Egyptian concerns. It is plausible that despite Nasser's public rhetoric Egyptian leaders did not perceive external threats as warranting nuclear weapons.[39] However, that possibility would undermine the basic neorealist argument that adversaries' efforts to acquire nuclear weapons drives "reactive proliferation." Third, Egypt lacked conventional superiority over some of those potential adversaries but conventional capabilities seemed to have played a far lesser role than neorealism would predict. Egypt considered nuclear weapons under Nasser when its conventional gap with Israel was the narrowest, and abandoned nuclear aspirations under Sadat when the gap had widened. Egypt could muster only progressively declining conventional deterrence. Fourth, Egypt inhabited a multipolar environment that neorealist predictions clearly associate with nuclearization. Fifth, Egypt lacked a hegemon explicitly providing a nuclear umbrella, except for vague Soviet commitments. Indeed, Egypt considered nuclear weapons when it enjoyed stronger external (Soviet) security guarantees but abandoned them in their absence (the US never provided Egypt with equivalent guarantees). Sixth, Egypt was no common regional power but rather the leading Arab power expected to operate according to the logic of power balances.

Thus, Egypt's nuclear renunciation despite all these circumstances posits a serious anomaly for neorealism in a region prone to nuclearization, where at least seven states sought to acquire nuclear weapons' capabilities.[40] Changes in

[39] Even Egypt's foremost advocate of nuclear weapons, Nasser's advisor Mohamed Hasanayn Haykal, dismissed the premise that Israel's nuclear designs posed an offensive threat to Egypt, acknowledging that "Israel has nuclear weapons but will not use them unless she finds herself being strangled." Shai Feldman, *Israeli Nuclear Deterrence: A Strategy for the 1980s* (New York: Columbia University Press, 1982), p. 87.

[40] Bruce Riedel and Gary Samore, *Managing Nuclear Proliferation in the Middle East* (Washington, DC: Brookings Institution Press, 2008). Egypt could have relied on its latent capabilities to build nuclear weapons if it had chosen to, according to Meyer, *The Dynamics of Nuclear Proliferation*, p. 109. See also Shyam Bhatia,

relative power do not provide coherent accounts of Egypt's rejection of "reactive proliferation" since the 1970s. Explaining why Egypt was not part of a nuclearizing trend among Middle Eastern contenders requires a consideration of alternative arguments driving nuclear decisions. Furthermore, Egypt's position regarding nuclear weapons shifted overtime. Egyptian expert Mohamed K. Said stated that "there is a lack of consensus among the Egyptian elites over the strategic value of acquiring nuclear weapons capabilities. In the 1960s, Egypt was firmly committed to the search for strategic nuclear balance with Israel. By the 1970s, however, Egypt had become less enthusiastic about the issue."[41] Gamal Abdul Nasser invested in it but Sadat and his successors renounced it; this compels a more serious understanding of these leaders' respective models of survival in power.

Both early efforts toward nuclearization by Nasser and the subsequent shift by Sadat were compatible with their respective models. Nasser's inward-looking coalition was acutely aware of its primary domestic and regional audiences, and a nuclear program provided a particularly valuable tool of domestic survival. Sadat favored nuclear transparency as part of a broader shift to substitute Nasser's model of inward-looking self reliance, hyper-nationalism, and military-technical prowess with Sadat's own blueprint of export-led economic growth, foreign investment, military conversion, and new relationships with international markets and institutions, all of which left no room for an ambiguous and expensive nuclear program. Sadat read international incentives differently from Nasser, sought to build alternative bases of support (domestic and foreign), and made extensive use of US economic, political, and military incentives. Sadat's initiatives—*infitah* (economic reform), growth, and Egypt-first policies—were compatible with NPT ratification in 1981. President Hosni Mubarak retained nuclear abstention despite strong domestic revisionist currents advocating nuclearization.[42]

Nuclear Rivals in the Middle East (New York: Routledge, 1988), p. 47. A 1963 proliferation estimate by Secretary of Defense Robert McNamara deemed Egypt's motivations to obtain nuclear weapons "moderate to high." See Peter Lavoy, "Predicting Nuclear Proliferation: A Declassified Documentary Record," *Strategic Insights*, Vol. 3, No. 1 (January 2004).

[41] Mohamed Kadry Said, "Security and Defense Dilemmas in the Middle East: The Nuclear Dimension," Pugwash Meeting No. 279, Vol. 2 (November 15–17, 2002) at http://www.pugwash.org.

[42] For a more detailed analysis of Egypt's nuclear trajectory until 2007 than is possible here, see chapter 10 in Solingen, *Nuclear Logics*.

As threats of Iran's nuclearization grew stronger, so did signs of possible revisions in Egypt's policy. Iran's nuclear defiance became an important factor in Egypt's statements. The relevant causal mechanisms, however, had far more to do with domestic pressures from within Egypt than with fears of an Iranian military offensive. The neorealist account is often invoked in this context. Yet for it to be convincing in explaining continuity and change in Egypt's nuclear policy it would have to provide persuasive evidence of the following:

(1) That Iran is a *strategic* threat to Egypt rather than a political one (which would provoke growing domestic pressures from pro-nuclear forces in Egypt bent on either countering, matching or emulating Iran's nuclear defiance);
(2) that distant Iran is a far more serious threat to Egypt's security than adjacent Israel has been, given that no dramatic steps were taken in decades, presumably because there was no perceived security threat until Iran's nuclearization;
(3) that shifts in relative power have driven such decisions independently of other causal mechanisms, including cited domestic considerations;
(4) that Egypt's approach to Iran's nuclearization is *not* connected to Sunni-Shi'a competition, a driver more embedded in constructed understandings of religion than in classical balance-of-power considerations; and
(5) that Egypt's changed nuclear behavior is not motivated by a dramatic deterioration of the non-proliferation regime in the aftermath of North Korea's test and Iran's violations. Without strong positive evidence that these five drivers account for a possible shift in Egypt's position in recent years, our belief in neorealist balance-of-power explanations as the commanding force behind Egypt's evolving nuclear behavior would be weakened. As argued, alleged signs of possible revisions in Egypt's policy may have far more to do with domestic pressures from resilient inward-looking constituencies that endured under Mubarak's regime as well as in its aftermath. Mubarak's Egypt never signed the Additional Protocol, and it accompanied demands for a Middle East nuclear weapons free zone with veiled threats to develop its own nuclear weapons if others did not give them up. Strong inward-looking constituencies amidst flawed economic reforms (including corrupt privatization schemes and continued protectionism of vast economic sectors under military control, among others) precluded Egypt's genuine transformation into a strong internationalizing political economy. Aside from segments of the military, inward-looking

constituencies included former Nasserites, the Muslim Brotherhood, and those employed in state enterprises and import-substituting, uncompetitive industries.[43] Iran's skillful use of the Palestinian issue, including its shoring up of the militant Hamas and Hizbullah groups, gained it support among segments of the Egyptian public.[44] Israel's intransigent coalition under Benjamin Netanyahu, unable or unwilling to take more effective steps toward a settlement with the Palestinian Authority, helped maintain populist support for Iran as Israel's main adversary even in the aftermath of harsh domestic repression by Iran's hardliners, particularly since 2009.

Mubarak's Foreign Minister Ahmed Abul-Gheit, among others, pointed to financial and other Iranian support for Egypt's Islamist militant groups, suggesting that without the elimination of such support Egypt could not re-establish diplomatic relations with Iran.[45] This and other statements reveal significant concern with the spread of Shi'ism within Egypt.[46] Furthermore, in the words of an *Al-Ahram* analyst, "to Iran, Egypt's efforts to bring about peace in the region, and its encouragement of the Palestinians and other Arab countries to follow its path with Israel, threaten to deprive Tehran of the single most potent regional issue that it can exploit to further its radical agenda."[47] Beyond this issue, however, the article acknowledges that Iran's intervention in Egypt, as well as in Lebanon, Syria, Gaza, Yemen, Iraq and the Gulf, raises the specter of a "nuclear-armed, non-Arab, and Shi'a superpower expanding its influence further in the region." Iran's Islamic regime had also supported

[43] The Muslim Brotherhood does not believe in the concept of the state but rather in the Islamic *ummah* (nation), and the movement's Supreme Guide once said, "To hell with Egypt!" see Tariq Alhomayed, "Who's Interested in Arabism Today?", *Al-Sharq Al-Awsat* (October 29, 2010) at http://www.aawsat.com/english/news.asp?section=2&id=22808.

[44] Dina Ezzat, "Not Good, Not Bad," *Al-Ahram Weekly*, No. 947 (Cairo: 14–20 May 2009) at http://weekly.ahram.org.eg/2009/947/eg1.htm (last accessed October 17, 2010).

[45] Dina Ezzat, "More than dialogue needed," *Al-Ahram Weekly*, No. 865 (Cairo: 4–10 October 2007)at http://weekly.ahram.org.eg/2007/865/eg2.htm (last accessed October 17, 2010).

[46] Salah Hemeid, "Rapprochement with Iran," *Al-Ahram Weekly*, No. 977 (Cairo: 17–23 December 2009) at http://weekly.ahram.org.eg/2009/977/re52.htm (last accessed October 17, 2010).

[47] Ibid.

the erection of memorials in Tehran praising Sadat's assassin, an Islamist militant named Khaled El-Islambouli.[48]

Some opposition members in Egypt's Shura Council (consultative upper house), supported by independents, called for Egypt to develop a long-term nuclear weapons program in response to both Israel and Iran.[49] Nagui El-Shehabi, appointed member of the Shura and chairman of the Geel (Generation) opposition party, "the Geel" (or the Democratic Generation) opposition party, proposed the same in a Council discussion of Egypt's nuclear power program. Osama Shaltout, another appointed Shura member and chairman of the Social Justice Party, insisted that Egypt should revoke its NPT membership. Rifaat El-Said, chairman of the leftist Tagammu Party, cited Egypt as the first Middle Eastern state to initiate a nuclear energy program in the mid-1950s, adding that President Nasser was a visionary in seeking nuclear weapons against Israel.[50] Mamdouh Qenawi, chairman of the Free Constitutional Party, accused Israel of refusing to sign the NPT. Egyptian opposition members also criticized the government for reports that US and Israeli ships designed to deter Iran had used Egyptian territorial waters.

Fueling ambitions to enhance Iran's position among domestic constituencies in Egypt (and in other Arab countries), Ahmadinejad offered to restore diplomatic relations with Egypt. Yet concerns with Iranian meddling in Egypt's domestic affairs precluded normalization under Mubarak. Egyptian officials publicly supported a diplomatic solution to the conflict over Iran's nuclear program, but Foreign Minister Abul-Gheit also expressed support for "international efforts to prevent Iran from developing nuclear weapons."[51] *Daily News Egypt* reported that US-allied Arab nations, mainly headed by

[48] "Iran makes new bid for improving ties with Arabs," *Daily News Egypt*, (Cairo: December 23, 2009).

[49] Gamal Essam El-Din, "No need for nukes," *Al-Ahram Weekly*, No. 977 (Cairo: 17–23 December 2009) at http://weekly.ahram.org.eg/2009/977/fr1.htm (last accessed October 17, 2010).

[50] The claim that Egypt, not Israel, was first to seek nuclear weapons in the region finds support from other Egyptian sources. See Solingen, *Nuclear Logics*, chapter 10. See also the statement by the Chairman of the Shura Council's Industry Committee, Mohamed Farid Khamis: "Do not forget that Egypt was the first in the region to have a group of trusted nuclear experts," quoted in Gamal Essam El-Din, "No need for nukes."

[51] "Tense Egypt-Hamas relations take a turn for the worst," *Daily News Egypt* (Cairo: December 4, 2008).

Sunni Muslims, were deeply concerned with the rising power of Shi'a and Persian Iran; with Iran's nuclear program; with the potential negative effects of confronting Iran militarily; and no less with the possibility of US concessions to Iran.[52] An Egyptian analyst argued that Egyptians believed Western countries had grown more impatient with Iran's refusal to provide concrete guarantees that it was not pursuing nuclear weapons. Hence, the analysis stressed tougher economic sanctions and even an air strike against Iran's nuclear facilities, concluding with a quote from analyst Samir Ghattas, "The US bombing of Iraq in 1991 set the stage for the Madrid peace conference later in that year...so would an air strike against Iranian nuclear facilities prepare the region for another round of peace?"[53]

Mustafa El-Feki, who chaired the parliament's Foreign Relations Committee, described an Iranian "tendency for hegemony" and "Persian national aspirations."[54] The Minister of State for Military Production, Sayyed Al-Meshal, argued that Iran's military capabilities represent not a threat to Israel but a definite threat to Arab countries, interfering in their domestic affairs. Addressing the Egyptian Syndicate of Journalists, Al-Meshal emphasized that a nuclear Iran could trigger a European response to counter it.[55] Mahmoud Barakat, former head of the Arab Institute for Atomic Energy, rejected the view that Iran's nuclear program was a threat to Egypt's national security, arguing that "Egyptian national security comes from inside Egypt. Maintaining democracy and working hard to produce our needs are what will furnish us with security."[56] And Mubarak argued that Israel's concern with Iran was radically different from Egypt's. Whereas Egypt, he said, was concerned with Iran's "regional hegemony," Israel was worried about Iran's nuclear capabilities, adding that "if we talk of nuclear capabilities, we should talk of [Israel's] as well."[57] Medhat Hammad, editor of the *Iranian Strategic Report*, urged Egypt to counter Iranian cultural and religious influence by establishing a satellite

[52] "Iran makes new bid for improving ties with Arabs," *Daily News Egypt*.

[53] Saif Nasrawi, "Eyeing Iran, Egypt corners Hamas," *Al-Masry Al-Youm* (February 3, 2010) at http://www.almasryalyoum.com/en/node/14339 (last accessed October 14, 2010).

[54] Hemeid, "Rapprochement with Iran."

[55] http://www.albawaba.com/en/main/318089/&searchWords=Al-Meshal

[56] Doaa El-Bey, "The case for multi-tasking," *Al-Ahram Weekly*, No. 848 (Cairo: June 7–13, 2007) at http://weekly.ahram.org.eg/2007/848/eg8.htm (last accessed October 17, 2010).

[57] Ezzat, "Not Good, Not Bad."

channel in Farsi, targeting 40 million speakers, and building Egyptian cultural centers as well as reestablishing diplomatic relations with Iran. Nabil Othman, former chairman of the State Information Service, urged a strong economy as a means to unite Egyptians and enhance Egypt's status in the region.

Despite concerns over perceived Iranian interference in Egypt (and also in Iraq, Lebanon, Yemen and Palestine) through Hezbollah militants, and perhaps as a nod to domestic constituencies friendly to Iran, the Mubarak regime agreed to restart direct air flights to and from Tehran, which had not operated for about 30 years since relations were severed in 1979. Iran also reported an agreement to export 25,000 tons of petrochemical products to Egypt even as Foreign Minister Abul-Gheit reiterated Egypt's commitment to defend its brethren in the Gulf against any threats to their security, without mentioning Iran by name. Official responses to the Iranian conundrum also emphasized Egypt's peaceful nuclear activities. Moufid Shehab, former Minister of State for Legal and Parliamentary Affairs, cited Mubarak's statement that Egypt's nuclear program would be fully transparent and that Egypt's international efforts to bring Israel into nuclear disarmament would continue.[58] Countering statements by the opposition cited earlier, the former chairman of the Shura Council and secretary-general of the ruling National Democratic Party (NDP) Safwat El-Sherif reiterated that "Egypt has no need for a nuclear armament programme."[59] Hassan Younis, former Minister of Electricity and Energy, affirmed IAEA inspectors' full access to Egypt's nuclear program. Egyptian officials also denied an IAEA report that between 2007 and 2008 Egypt had used the Inshas reactor to produce small amounts of 20 percent-enriched uranium.[60]

In an effort to respond to a pincer movement emanating from the interaction between external and domestic pressures discussed thus far, Egypt under Mubarak led regional efforts to advance a nuclear weapons free zone that

[58] El-Din, "No need for nukes."

[59] Ibid.

[60] A secret report by Egypt's atomic agency argued that Egypt's interests could be harmed by an IAEA investigation of the origin of highly enriched uranium particles found at the Inshas Nuclear Research Center in 2007 and 2008 through inspections. The Egyptian report suggested that the IAEA Director General, Yukiya Amano, might not take as "lenient" a stance as Egyptian-born Mohamed ElBaradei had. See Global Security Newswire (June 10, 2010) at http://gsn.nti.org/gsn/nw_20101115_6940.php. The report was deemed a forgery by Egyptian Ambassador to the IAEA, Ehab Fawzy.

would force Israel to join the NPT and presumably prevent Iran from developing nuclear weapons. One triumph was the final document of the May 2010 NPT Review Conference which called for convening of a weapons of mass destruction free zone (WMDFZ) in the Middle East in 2012. However, an Egyptian-sponsored resolution at the 2010 IAEA annual conference, calling specifically for Israel to join the NPT, suffered a setback. As Cortright and Vayrynen argue, "no progress on a nuclear weapons free zone in the Middle East will be possible without a broader peace settlement in the region. ... An agreement structured along the lines of the Arab initiative first proposed by Saudi Arabia in 2002 would be a *sine qua non* for any consideration of a nuclear weapons free zone in the region."[61]

Prior to the 2011 uprising, one (disputed) public opinion poll reported that 69 percent of Egyptians, and 57 percent throughout Egypt, Saudi Arabia, Morocco, Jordan, Lebanon and the United Arab Emirates, saw a nuclear-armed Iran as "better for the Middle East."[62] However, a highly authoritative poll (Pew Global Attitudes Survey) suggested otherwise.[63] Majorities or pluralities in four predominantly Muslim countries had unfavorable opinions of Iran: 66 percent in Egypt, 63 percent in Jordan, 60 percent in Lebanon and 58 percent in Turkey. Only a majority in Pakistan and Indonesia were found more favorable to Iran.[64] Among those who opposed Iran's acquisition of nuclear weapons, 55 percent of Egyptians expressed support for military force (with 53 percent in Jordan and 44 percent in Lebanon). A German Marshall Fund poll found that 25 percent in Turkey were willing to accept a nuclear Iran, somewhat at odds with the poll claiming wide support for Iran throughout the Middle East. Opposition to Iran's nuclear program was very strong

[61] David Cortright And Raimo Väyrynen, "A Policy Agenda for Enhancing Security Without Nuclear Weapons," *Adelphi Papers*, Vol. 49, No. 410 (October 2009).

[62] Shibley Telhami, "2010 Arabic Public Opinion Poll," The Brookings Institution (August 5, 2010 at http://www.brookings.edu/~/media/Files/rc/reports/2010/08_arab_opinion_poll_telhami/08_arab_opinion_poll_telhami.pdf.

[63] "Muslim Disappointment: Obama More Popular Abroad Than at Home, Global Image of U.S. Continues to Benefit," The Pew Global Attitudes Project (June 17, 2010) at http://pewglobal.org/files/pdf/Pew-Global-Attitudes-Spring-2010-Report.pdf.

[64] About 58 percent favored Iran's pursuit of nuclear weapons in Pakistan, rising to 91 percent among Lebanese Shi'a. See Bruce Stokes, "Iran's nuclear program: the public speaks," *The Daily Star* (Lebanon: October 4, 2010) at http://www.dailystar.com.lb/article.asp?edition_id=1&categ_id=119967# axzz12gcGCLLP.

across most countries surveyed by the Pew poll, with significant support for tougher economic sanctions. A 2010 BBC-sponsored poll revealed that unfavorable attitudes toward Iran in Egypt (36 percent) overwhelmed positive views of Iran (27 percent).[65] The usual caveats about polls—which are highly contingent upon the questions asked and can be highly volatile in response to events—apply.

The revolutionary upheaval of early 2011 makes it particularly challenging to assess the extent of internal support for Iran's regime and policies within Egypt. Not every criticism of the former authoritarian regime necessarily implies a vote for Iran's policies. Ambassador Menha Bakhoum, spokeswoman for the Foreign Ministry, expressed that "All the world has diplomatic relations with Iran with the exception of the United States and Israel," and that Egypt looks at Iran "as a neighbor in the region" that Egypt should have normal relations with. "Iran is not perceived as an enemy as it was under the previous regime, and it is not perceived as a friend," she argued. Noting that Egypt broke off relations with the Islamic Republic after its 1979 revolution, she recalled that limited relations reopened in 1991 at the chargé d'affaires level, and hence normalizing relations was more of an elevation than a reopening. The new Egyptian Foreign Minister Nabil al-Arabi expressed readiness to enhance ties with Iran as part of Egypt's redefinition of its foreign policy. Al-Arabi is also the new Secretary General of the Arab League.

After approval from Egypt's new caretaker government, Iranian ships passed through the Suez Canal en route to Syria for the first time in three decades. The possibility of closer relations between Egypt and Iran is a concern for some Gulf countries. However, even some Egyptian supporters of normalization with Iran have tried to reassure them that the security of the Gulf Arab region is an integral part of Egypt's security.[66] More broadly, the irony of a democratizing Arab region embracing an increasingly more repressive Iranian regime is not lost on regime supporters or opponents in Iran, or on the Egyptian public for that matter.[67] Egypt's trailblazing effect on the rest

[65] "Global Views of United States Improve While Other Countries Decline," World Public Opinion at http://www.worldpublicopinion.org/pipa/pipa/pdf/apr10/BBCViews_Apr10_rpt.pdf (last accessed October 17, 2010).

[66] Aymen El-Emir, "Israel's worst fear," *Al-Ahram Weekly*, No. 1046 (Cairo: May 5–11, 2011) at http://weekly.ahram.org.eg/2011/1046/op6.htm.

[67] Amani Maged, "Coup in Coup Square," *Al-Ahram Weekly* No. 1037 (Cairo: March 3–9, 2011) at http://weekly.ahram.org.eg/2011/1037/re113.htm.

of the region; Syria teetering at the verge of deeper internal conflagration;[68] hazardous Iranian intrusion in the domestic politics of Arab states in the Persian Gulf; and Turkey's tainted status given its leaders' close connections with repressive Middle East leaders, are all ingredients for a challenging future facing Iran's repressive hardliners and their inward-looking plutocratic allies. Deceitful support for democratic forces *elsewhere* in the region has done nothing to diminish this challenge so far.

It is impossible to predict what kinds of regimes and political-economy models will take root in Egypt, Tunisia, Yemen, Syria and beyond in the years to come. As an astute scholar of the Middle East argues, just as in Iran in 1979, demonstrators in 2011 were united in their opposition to authoritarian regimes but also had wildly differing goals.[69] Demands for jobs, economic development, low food prices and other welfare objectives will clearly dominate. Competing coalitions will vie for alternative ways to meet those growing demands, either through deepening integration in the global economy or through more inward-looking strategies. The nature of the emerging political economy models will weigh heavily not only on the way Egypt, Iran and the rest of the region respond to these challenges but also on the nature of their nuclear postures in a new Middle East.

[68] A popular chant in Syria's popular protest movement is "Neither Iran, nor Hezbollah," unsurprisingly, given the presence of Iran's repressive apparatus within Syria itself.

[69] Timur Kuran, "The Politics of Revolutionary Surprise," *Project Syndicate* (February 8, 2011) at http://www.project-syndicate.org/commentary/kuran1/English.

3

Leadership Transitions and Nuclear Futures in Egypt and Libya

Maria Rost Rublee

Questions over the potential military uses of Iran's nuclear program have done far more than focus attention on Tehran's capabilities and intentions. Now, analysts raise the specter of a nuclear arms race in the Middle East, as the region has suddenly displayed great interest in nuclear technology. In the past five years, thirteen Middle Eastern countries have announced new or re-launched civilian nuclear programs.[1] While these programs are civilian in nature, some analysts are concerned that they could be intended as backdoors into military nuclear capability to match Iran.[2] However, while these argu-

I would like to thank Caitlin Merriman for invaluable research assistance.

[1] "Arab nations advance quest for atomic power," *Global Security Newswire* (July 15, 2010) at http://www.globalsecuritynewswire.org/gsn/nw_20100715_5004.php.

[2] For an in-depth analysis of neighboring countries' responses to Iran's nuclear program, see Dalia Dassa Kaye and Frederic M. Wehrey, "A Nuclear Iran: The Reactions of Neighbors," *Survival, Vol.* 49, No. 2 (2007): pp. 111–28. See also James A. Russell, "A Tipping Point Realized? Nuclear Proliferation in the Persian Gulf and Middle East," *Contemporary Security Policy* , Vol. 29, No. 3 (2008): pp. 521–37; and Richard

ments have a certain logic, they impute the weight of explanatory power to regional dynamics, and may not recognize the importance of other factors in the creation of nuclear programs—impetuses both for and against the creation of nuclear weapons.

This chapter proposes to counterbalance the focus on regional strategic dynamics with an analysis of the importance of individual-level decision making in two key countries in the Middle East: Egypt and Libya. The bulk of scholarly work on nuclear proliferation uses realism, which seeks to understand nuclear politics through systemic security drivers.[3] This trend has slowly turned toward a greater theoretical focus on other variables, including economic conditions, domestic coalitions, normative concerns, institutional constraints, and even supply and demand.[4] Few theoretically informed works,

---

Weitz, "Gulf Cooperation Council Moves Forward with Nuclear Energy Plans," *WMD Insights* (April 2007) at http://www.wmdinsights.com/I14/I14_ME3_GCCMovesForward.htm.

[3] See Kenneth N. Waltz, "Nuclear Myths and Political Realities," *American Political Science Review*, Vol. 84, No. 3 (September 1990): pp. 731–45; Kenneth Waltz, "The Spread of Nuclear Weapons: Why More May Be Better," Adelphi Papers , No. 171 (London: International Institute for Strategic Studies, 1981); Benjamin Frankel, "The Brooding Shadow: Systemic Incentives and Nuclear Weapons Proliferation," in Zachary Davis and Benjamin Frankel (eds), *The Proliferation Puzzle: Why Nuclear Weapons Spread* (London: Frank Cass, 1993); Stephen Van Evera, "Primed for Peace: Europe after the Cold War," *International Security*, Vol. 15, No. 3 (Winter 1990–1).

[4] For analysis using domestic politics and political economy, see Etel Solingen, "Domestic Sources of International Regimes: The Evolution of Nuclear Ambiguity in the Middle East," *International Studies Quarterly*, Vol. 38, No. 4 (June 1994); Etel Solingen, "The Political Economy of Nuclear Restraint," *International Security*, Vol. 19, No. 2 (Fall 1994); Etel Solingen, "The New Multilateralism and Nonproliferation: Bringing Domestic Politics In," *Global Governance*, Vol. 1, No. 2 (May-August 1995). Solingen's most recent analysis of nuclear decision making is found in her *Nuclear Logics: Contrasting Paths in East Asia and the Middle East* (Princeton, NJ: Princeton University Press, 2007). For early academic work that raised the importance of norms in nuclear politics, see, for example, Steven Flank, "Nonproliferation Policy: A Quintet for Two Violas?", *Nonproliferation Review*, Vol. 1, No. 3 (1994): pp. 71–81; Scott Sagan, "Why Do States Build Nuclear Weapons?: Three Models in Search of a Bomb," *International Security*, Vol. 21, No. 3 (Winter 1996–7): pp. 54–86; Tanya Ogilvie-White, "Is There a Theory of Nuclear Proliferation? An Analysis of the Current Debate," *Nonproliferation Review*, Vol. 4, No. 1 (Fall 1996):

however, have specifically spotlighted the role of individuals in nuclear decision-making; the notable exception is Jacques Hymans' *The Psychology of Nuclear Weapons*, which posits the psychological characteristics of decision-makers as a key variable.[5] The lack of attention to individuals may be due in part to the dominance of realism, but also to the more general neglect of the first image in international relations.

I argue that in nuclear politics, it may be time for a renaissance of the individual level of analysis—to bring the individual back in—because of the extraordinary domestic changes that occurred in the Arab Spring of 2011. In both Egypt and Libya, domestic upheavals have dramatically affected nuclear politics, and nuclear politics have the potential to affect domestic politics. The paucity of strictly systemic level analysis comes into full light when considering the dramatic differences in nuclear decision making that can be predicted by envisioning the variety of potential leadership transitions in each country. The goal is not argue that system or state level attributes are not important, but rather that the individual level must also be examined for an honest look at nuclear futures in Egypt and Libya. In short, this chapter argues that to understand Egyptian and Libyan nuclear prospects, we must wrestle with the question of who will lead these countries. This chapter will first detail a social-psychological framework through which individual nuclear decision making can be understood, and then discuss succession and nuclear politics more generally. Then the chapter will turn to individual examinations of both Egypt and Libya.

pp. 43–60. For work that has focused directly on identities and norms in nuclear decision making, see Maria Rost Rublee, *Nonproliferation Norms: Why States Choose Nuclear Restraint* (Athens, GA: University of Georgia, 2009); Harald Müller, "German National Identity and WMD Proliferation," *Nonproliferation Review*, Vol. 10, No. 2 (Summer 2003): pp. 1–20; Jacques Hymans, *The Psychology of Nuclear Weapons: Identity, Emotions and Foreign Policy* (Cambridge University Press, 2006); Jacques Hymans, "Isotopes and Identity: Australia and the Nuclear Weapons Option, 1949–99," *Nonproliferation Review*, Vol. 7, No. 1 (Spring 2000): pp. 1–23; and Nina Tannenwald, *The Nuclear Taboo: The United States and the Non-Use of Nuclear Weapons Since 1945* (Cambridge University Press, 2007).

5 Jacques E.C. Hymans, *The Psychology of Nuclear Proliferation: Identity, Emotions and Foreign Policy* (Cambridge University Press, 2006). See also William J. Long and Suzette R. Grillot, "Ideas, Beliefs, and Nuclear Policies: The Cases of South Africa and Ukraine," *Nonproliferation Review*, Vol. 7, No. 1 (Spring 2000): pp. 24–40.

Individuals and the social psychology of nuclear decision making

While neorealism explicitly minimizes the influence of individuals in international security, constructivism focuses on the social construction of politics, opening a door to the importance of individual beliefs and ideas.[6] Of course, constructivism does not mandate individual-level explanations; for example, constructivist scholar Alexander Wendt concentrates on the system level.[7] However, once ideas are acknowledged as an important explanatory variable, analysis of individuals makes sense, given the possibility of elites influencing decision making, especially in arenas where policymaking is secretive and open to only a small circle, such as nuclear politics.[8]

I propose to use a social-psychological framework to understand elite orientations toward nuclear proliferation and non-proliferation, and in particular, to understand the different types of motivation for non-proliferation. With regard to the choice of whether or not to develop nuclear weapons, typically analysts divide countries (or individuals) into two camps: rejection or compliance with the non-proliferation norm. A binary choice, however, is too crude to reflect the realities of nuclear decision making. As some have noted, rejection may not lead to outright pursuit of nuclear weapons; states can engage in nuclear hedging or make other decisions that allow them to get closer to a nuclear capability without having to cross the red line.[9] However, the "compliance" choice also needs to be more nuanced. I argue that compliance with the non-proliferation norm can best be understood as three separate outcomes: persuasion, identification, or conformity.

Persuasion is best characterized as both public and private acceptance of the norm: "I now see that X is better than Y." In other words, persuasion results from the genuine transformation of preferences. Conformity is public acceptance with private rejection: "I think Y is correct (or I like Y better), but since everyone else says X, I will do X so I don't rock the boat."[10] Conformity results

[6] The neorealist case against the individual level is stated most clearly in Kenneth Waltz, *Theory of International Politics* (Reading, MA: Addison-Wesley, 1979).

[7] See, for example, Alexander Wendt, "Collective identity formation and the international state," *American Political Science Review*, Vol. 88, No. 2 (1994): pp. 384–95.

[8] See for example, Jeffrey Checkel, "Ideas, Institutions, and the Gorbachev Foreign Policy Revolution," *World Politics*, Vol. 54, No. 2 (1993): pp. 271–300.

[9] Ariel Levite, "Never Say Never Again: Nuclear Reversal Revisted," *International Security*, Vol. 27, No. 3 (Winter 2002–3): pp. 59–88.

[10] Alastair Iain Johnston, "Treating International Institutions as Social Environments," *International Studies Quarterly*, Vol. 45, No. 4 (2001): pp. 498–9.

from the desire to maximize social benefits and/or minimize social costs, without a change in underlying preferences. The third outcome is identification, which is acceptance based on the desire or habit of following the actions of an important other: "Actor A says X, and I like (or want to be like) A, so I will say X also." In this case, there is little internal reflection on the actual norm; the focus is on the relationship with the desired actor.[11]

Breaking down compliance into three separate outcomes is important for policy reasons. Models are theoretical constructs that help us understand reality; they do not chart it out in full detail. However, mechanisms that produce original policy results *should be* included in models. An example from the non-proliferation arena will illustrate. Under George W. Bush, many argue that the United States backed away from some of its obligations under the Nuclear Non-Proliferation Treaty (NPT): rejection of discussion of disarmament obligations, public declaration of the decision to continue designing and computer-testing new nuclear weapons, and public threats to use nuclear weapons against non-nuclear weapons states. If an ally of the United States (call this state Ally X) had initially followed the US lead on nuclear non-proliferation due to persuasion, Ally X would remain persuaded, and thus would likely express disappointment, as well as encourage the United States to get back on course. If, however, another ally's behavior was based on identification with the United States (call this state Ally Y), then which United States would it identify with: the US of action or the US of rhetoric?[12]

In addition, because in reality states are not unitary actors, the distinctions between persuasion, conformity and identification likely play out in domestic politics. In fact, it is likely that each of the influence outcomes could be represented by some segment of society interested in nuclear policy. Examples would be NGOs and activists being "persuaded" that nuclear weapons are detrimental to state prestige and identity, policy wonks in the diplomatic core "identifying" with their Western allies, and members of the military bowing to "social conformity." In each case, the behavior is the same: nuclear forbearance. The reasons behind the actions are different, however, and material or social changes could lead to behavioral changes. A short narrative of how different domestic factions might play out in Ally X and Ally Y will illustrate.

[11] For a more in-depth discussion of these concepts, see Rublee, *Nonproliferation Norms*, pp. 17–21.

[12] Note that identification can be based on the desire to receive approval and acceptance from the in-group, in which case the US's continuing rhetoric against nuclear acquisition would likely keep Ally Y from changing policy.

This narrative is not meant to describe any two countries, but rather simply to highlight how differently influenced groups might stand with regard to nuclear policy, and how they might react to US behavior.

In the case of Ally X and Ally Y (both confronted by the US's changing behavior with regard to the nuclear non-proliferation regime), Ally X's nuclear policy could be supported by a coalition of civil servants in a bureaucracy that has supported the NPT for many years, political appointees who believe in non-proliferation, and anti-nuclear activists with embedded ties to the policymaking apparatus—all of whom are "persuaded." The United States' current actions would probably inspire disappointment, resentment, disgust—but a change of heart is not likely because these actors are genuinely persuaded of the merits of non-proliferation. However, other elements in that government and state—those who support non-proliferation because of identification or social conformity—will likely have a different reaction. Those who believe their state should forgo nuclear weapons due to the negative diplomatic effects any other position would have might rethink their position in light of the US's stance, as well as North Korea's ability to keep its nuclear weapons despite great power protests. In the short-run, it is not likely that the state's behavior would change, but in the long-run, those persuaded may change their minds or may lose ground to growing ranks of those who disagree. In the case of Ally Y, where nuclear policy is guided by identification with the United States, confusion is likely to result, based on a gap between US rhetoric and US actions. How do you behave when the one you have patterned yourself after starts to do something different from what they have said all along? Depending on the strength of the persuaded and conforming segments, and the result of any internal struggle between them, the state could move more definitively against nuclear acquisition or could move toward exploring the nuclear option.

In the same way, these distinctions are important with individual decision-makers. As the next section will detail, nuclear decisions are often made by a few key elite people, whose personalities, preferences and ideas can dramatically shape their perceptions and thus their behavior.

Leadership transitions and nuclear decision making

The literature on nuclear decision making mostly avoids the issue of the importance of individuals. Some policy-oriented work has acknowledged the importance of individual leaders in nuclear decisions. For example, Mitchell

Reiss concludes after his review of nine case studies that the most significant factor influencing nuclear decisions is the "quality of political leadership."[13] However, as Hymans notes,

> ... focus on individual leaders is unusual in the social-scientific literature on proliferation and non-proliferation. Indeed, most authors on the subject hardly even bother to ask the question of how leaders come to desire nuclear weapons. Instead, they simply adopt a tragic sensibility, viewing nuclear weapons as a system of a fallen humanity's raw quest for power.[14]

While much of international relations scholarship has avoided focus on the individual level, this lack of attention is particularly curious in the case of nuclear decision making. Again, Hymans notes,

> To go nuclear is an ideal-typical "big decision." In light of this, the standard menu of "security," "prestige," or "domestic political" motivations for foreign policy choice is insufficient. The consequences of going nuclear are simply too vast to allow for a reasonable cost-benefit calculation. To be sure, various voices in society may sound strong pro- or anti-bomb notes; but the responsibility for choosing wisely is much heavier for the top leader into whose hands the ultimate choice actually falls.[15]

The only other major work to focus on individual preferences as a critical role in nuclear decision making is a PhD dissertation by K.P. O'Reilly at the University of South Carolina, which argues that an "examination of the role leaders' perceptions play in proliferation decisions re-emphasizes the need for the scholarship to equally address the "opportunity" and "willingness" driving proliferation outcomes."[16]

Examination of cases of nuclear proliferation, non-proliferation and disarmament suggest that individuals are a necessary (though not sufficient) part of the puzzle. South Africa, the only country to create and then dismantle a nuclear weapons program, illustrates the importance of individuals in both proliferation and disarmament. P.W. Botha, Defense Minister when Pretoria began to lose Western support, was the key individual in seeing the South African bomb to completion. The South African leadership identified strongly

[13] Mitchell Reiss, *Bridled Ambition: Why Countries Constrain Their Nuclear Capabilities* (Washington, DC: Woodrow Wilson Center Press, 1995): p. 329.

[14] Hymans, *The Psychology of Nuclear Proliferation*, p. 1.

[15] Ibid., pp. 10–11.

[16] K.P. O'Reilly, "Beliefs And The Bomb: The Role of Leaders' World Views and Strategic Interactions on Proliferation Decisions," PhD dissertation (University of South Carolina, 2009) p. 244.

with the West and felt betrayed and abandoned by the West's rejection of them due to apartheid, and Botha especially felt the isolation. When Botha became Prime Minster in 1978, he quickly moved forward on nuclear weapons (while previous administrations did consider a bomb program, the program went into full swing under him). He declared that South Africa was locked in a "total war" that required extreme tactics. This mentality is illustrated by Botha's tendency to see his regime locked in an apocalyptic struggle, in a "world of good and evil, chosen people and heathens, redemption and fall." Nuclear weapons, never designed to be used militarily, gave Botha "emotional comfort" in face of Western rejection.[17] While Botha did perceive a security need for nuclear weapons, that was because of his paranoia—nuclear weapons were not needed to protect South African security. Pretoria faced no nuclear threat and could handle any conventional attacks with its superior conventional forces.

Just as Botha was a critical element in the South African acquisition of the bomb, so was F.W. de Klerk in its dismantlement. De Klerk had resolved to end the nuclear program before he even came into office. Immediately upon taking office in September 1989, he ordered an examination of the nuclear program, including bomb dismantlement and a timetable on how quickly South Africa could join the NPT. Entire dismantlement was completed by July 1990 and South Africa joined the NPT one year later. De Klerk was strongly motivated by a goal of ending his country's extreme isolation and bringing South Africa back into the community of nations; first and foremost, this meant ending apartheid, but it also meant giving up the nuclear program and joining the NPT. Some argue that de Klerk simply did not want the African National Congress (ANC) to gain control of nuclear weapons. That misses the larger point: why did de Klerk want to end apartheid in the first place? De Klerk's desire to fundamentally change South Africa and reintegrate with the world led him to end both apartheid and the nuclear weapons program.

Some argue that withdrawal of communist forces from South Africa clearly affected South Africa's decision making, and that is logical. However, it did not necessitate South African disarmament. After the 1988 Namibia cease-fire agreement, South Africa announced it had the capacity to make nuclear weap-

[17] Terence McNamee. "Denuclearizing a Regime: What South Africa's Nuclear Rollback Might Tell Us About Iran", Defense & Security Analysis, Vol. 22, No. 3 (September 2006): p. 333.

ons if needed and began preparations for a nuclear test as "insurance policy" in case the negotiated settlement broke down. Botha likely would have continued the program; in the mid-1980s, a new nuclear weapons facility was constructed, costing US$10 million, and it was completed just before de Klerk took office. De Klerk himself has argued that the end of the Cold War reduced the threats to the country and thus allowed him to disarm. However, numerous analysts, including Sagan and Reiss, have noted that this sounds like post-hoc rationalization. The Berlin Wall did not come down until November 1989, months after de Klerk made the decision to disarm (in fact, de Klerk wanted to end the program before he even took office). The defense community was strongly opposed to giving up the nuclear weapons program; in other words, most did not agree with de Klerk's assessment that South Africa's security environment had improved enough to make such drastic changes. De Klerk was concerned that right-wing opposition to his policies could lead to violence. Therefore, a security justification for what he already wanted to do may have made disarmament easier for the conservatives to swallow.

Personal leadership was also found to be important in both proliferation and non-proliferation decisions in Egypt, Indonesia, Romania and Yugoslavia. A National Defense University study argued,

> For all four cases, the importance of personal leadership cannot be overstated. In each case, the autocratic leadership saw a nuclear program as a way to achieve their larger visions both domestically and internationally. Nasser's drive to lead the Arab world—in particular in its struggle against Israel—was the "cardinal principle" of his foreign policy. Sukarno saw the nuclear bomb as a way to advance his new Indonesian revolution. Ceausescu was consumed with the need for independence, not just in foreign policy, but also in nuclear technology. Tito pursued nuclear weapons first to enhance his bid for strategic independence and then again in face of an Indian threat to his leadership of the non-aligned movement. Closely tied to the leadership factor is the desire for increased international standing and prestige. Indeed, all four states were led by leaders with significant egos. The pursuit of nuclear weapons was a means by which to demonstrate their personal and their country's greatness.[18]

For all four countries, removal of the main proponent of the nuclear program led to its end. This was particularly the case in Indonesia and Romania. As Robert M. Cornejo notes, "The idea of an Indonesian bomb had been so closely tied to Sukarno and his vision of a new Indonesian revolution that there was no real chance that Indonesia's new leader would pursue nuclear

[18] Rebecca Hersman and Robert Peters, *Nuclear U-Turns: Lessons from Historical Cases of Rollback* (Washington, DC: National Defense University, 2011), p. 104.

weapons."[19] In a similar vein, after Ceausescu was executed, many Romanian nuclear experts left the country, troubled by his heavy-handed push toward nuclear independence (reported to include slave labor); revulsion against Ceausescu's policy led the country to immediately close the nuclear weapons program.[20]

In particular, nuclear politics can be dramatically shaped by succession politics. North Korea is a case study in how nuclear weapons can ensure a regime's domestic survival. After the death of the founder of the Democratic People's Republic of Korea (DPRK), Kim Il Sung, his son and successor Kim Jong Il faced numerous challenges to his authority. At least two coup attempts were reported, and numerous high-level defections took place.[21] Kim Jong Il did not consolidate power until 1997 (three years after his father's death), and even then he was not appointed DPRK President (the position was abolished).[22] Analysts agree that the nuclear program has been used to shore up Kim Jong Il's position domestically, particularly with the military:

> When Kim Jong Il came to power after his father's death in 1994, there were many doubts that he would be accepted, especially by hardliners in the military. Given the DPRK's economic problems since Kim Jong Il came to power, there have been doubts regarding his leadership and his control of the military. The acquisition of nuclear weapons helps to bolster his position among the military and, in North Koreans' eyes, elevates the DPRK's prestige as a major player in Northeast Asia.[23]

Within North Korea, credit for spearheading the nuclear weapons program has gone to Kim Jong Il, and he has continued to use the nuclear weapons program to bolster his position domestically.[24] Inheriting his father's political position does not mean he also inherited the political authority that his father

[19] Robert Cornejo, "When Sukarno Sought the Bomb: Indonesian Nuclear Aspirations in the Mid-1960s," *Nonproliferation Review* (Summer 2000), p. 38.

[20] Hersman and Peters, *Nuclear U-Turns*, p. 97.

[21] "North Korea: Leadership Succession," *Global Security.org* at http://www.globalsecurity.org/military/world/dprk/leadership-succession.htm (last accessed December 12, 2009).

[22] Ibid.

[23] Terence Roehrig, "One Rogue State crisis at a time! The United States and North Korea's Nuclear Weapons Program," *World Affairs*, Vol. 165, No. 4 (April 2003): p. 168. See also Mun Suk Ahn, "After Kim Jong-Il," *Bulletin of the Atomic Scientists* (Chicago, IL: November 16, 2009) at http://www.thebulletin.org/web-edition/features/after-kim-jong-il. (last accessed December 5, 2009)

[24] Soo-Ho Lim, "Motives behind NK's Nuclear Weapons and Prospects for Denuclearization," *SERI Quarterly* , Vol. 2, No. 4 (October 2009), p. 120.

wielded. Indeed, because of the severe economic crises that the country has faced, Kim Jong Il has had to look elsewhere for sources of political legitimacy. He has done this in large part through the development of nuclear weapons and ballistic missiles, along with the confrontation with the United States that these programs create. By defying the international community, Kim Jong Il has been able to stir nationalist sentiment in his favor, and to strengthen his support among hardliners in the military, whom some call his most important constituency.[25] North Korea's nuclear test in May 2009, called "*juche* with a vengeance," has been interpreted as a means to strengthen Kim Jong Il's appointment of his son, Kim Jong-un, as his successor.[26] For politicians facing a succession crisis, particularly one that involves a dynastic succession from father to son, the story of how nuclear politics propelled a weak successor facing numerous coup attempts into a solidified position of leadership may be quite attractive.

To understand how leadership transitions and nuclear policy have influenced each other in the Middle East, the chapter next turns to individual case-study examinations of Egypt and Libya.

Leadership, transitions and nuclear decision making in Egypt

Egypt's nuclear history[27]

At the most critical periods in Egyptian nuclear history, individual elites played primary roles. Egypt's nuclear program began in 1954, the same year

[25] Reiss, *Bridled Ambition*, p. 99.

[26] Ibid. See also Victor D. Cha, "What Do They Really Want?: Obama's North Korea Conundrum," *Washington Quarterly*, Vol. 32, No. 4 (October 2009), pp. 125–6.

[27] This section on Egyptian nuclear history is drawn from research for my book, *Nonproliferation Norms*, chapter 4. Other important accounts of the Egyptian nuclear program include Jim Walsh, *Bombs Unbuilt: Power, Ideas and Institutions in International Politics*, PhD dissertation (Massachusetts Institute of Technology, 2001); Solingen, *Nuclear Logics*; Mark Fitzpatrick (ed.), *Nuclear Programmes in the Middle East: In the Shadow of Iran* (London: International Institute for Strategic Studies, 2008); Robert J. Einhorn, "Egypt: Frustrated But Still on a Non-Nuclear Course," in Kurt M. Campbell, Robert J. Einhorn and Mitchell B. Reiss (eds), *The Nuclear Tipping Point: Why States Reconsider Their Nuclear Choices* (Washington, DC: Brookings Institution, 2004), pp.43–82; and Gawdat Bahgat, "Nuclear Proliferation: Egypt," *Middle Eastern Studies*, Vol. 43, No. 3 (May 2007). Avner Cohen's authoritative work on Israel's nuclear program is also a rich source of information

that Gamal Abdul Nasser took power. Nuclear technology was seen as the wave of the future, and Nasser had a personal interest in it. He did not want Egypt to be left behind in the nuclear age, and this impression was reinforced by the power of nuclear medicine. As one former senior Egyptian science official said, "A favorite lady singer in Egypt developed thyroid problems, and she went to the Walter Reed hospital in the US to be treated with radiation therapy. After she came back, she told Nasser, and he was very impressed."[28] At this point, the nuclear program was meant to leave an opening for a military capability, but it was not designed for nuclear weapons.

Dramatic change occurred when the military dimension of Israel's reactor Dimona became public in 1960. Nasser responded with fiery rhetoric, and also directed the Atomic Energy Establishment (AEE) to look into military nuclear capabilities. However, the response from the AEE was mild; the main addition to its work after Nasser's directive was an emphasis on ensuring that the nuclear power reactors acquired were heavy water reactors, which were designed to provide the plutonium route to nuclear weapons.[29] By 1964, Egypt completed negotiations with Siemens, a West German company, for a natural uranium-fuelled, heavy water-moderated reactor.[30] Because this type of reactor had the potential to create fissionable material, the United States was concerned.[31] However, due to Egyptian anger over West Germany's tank sale to Israel, the deal was cancelled a year later.[32] So Cairo turned to Westinghouse, which had proposed a light water reactor. The reactor project never went forward, due at first to financing troubles and then to all nuclear programs being frozen as a result of Egypt's loss in the 1967 war. Nasser had also sought foreign assistance with the project, but these efforts also did not bear fruit.

The 1967 war with Israel ended, in all practical terms, Egypt's nuclear weapons program. Why did it fail? The reasons can be found largely in the indi-

---

on the Egyptian perspectives on nuclear weapons: see Avner Cohen, *Israel and the Bomb* (New York: Columbia University Press, 1998).

[28] Interview with former senior Egyptian science official, Cairo (March 2004).

[29] Ibid.

[30] Einhorn, "Egypt: Frustrated But Still On a Non-Nuclear Course," p. 46.

[31] "Attempts to Purchase 150 Megawatt Nuclear Reactor by the United Arab Republic," Secret Cable from United States Embassy in Egypt, Office of Air Attaché, to United States Air Force Chief of Staff (Digital National Security Archive, Item Number NP01098, January 17, 1965).

[32] Shyam Bhatia, *Nuclear Rivals in the Middle East* (New York: Routledge, 1988), p. 55.

vidual level of analysis. The program had serious project management problems, as well as infighting. However, strong political will could have overcome these issues. As one Egyptian security expert noted, "The program was very modest, and it suffered from a number of administrative and bureaucratic problems: engineering versus non-engineering, military versus non-military. But Nasser was not resolved to have a nuclear weapons program, so the problems never really got solved."[33] Indeed, Nasser's lack of commitment to the program is the best explanation for its failure. But this merely begs the question, why was Nasser uncommitted? There is some reason to believe he was concerned that nuclear weapons would strengthen the head of the AEE, Hedayat, and his mentor Hakim Amer, Egypt's Defense Minister. Others believe that personal appeals from US President John F. Kennedy influenced Nasser. Kennedy engaged Nasser in personal correspondence, in part to ask Egypt to abide by the spirit of the emerging non-proliferation regime. In both 1962 and 1963, Kennedy specifically asked for a pledge from Nasser that Egypt would remain non-nuclear.[34] At least one high-level Egyptian nuclear scientist believes these appeals influenced Nasser's decision making.[35]

In addition, Nasser was likely influenced by the emerging nuclear non-proliferation norm. Through 1965, the main result of high-level Arab discussions of Israel's nuclear potential seemed to be agreements to raise the issue at the United Nations and through other diplomatic channels.[36] Some Arab states criticized Nasser for his political and diplomatic approach to the Israeli nuclear issue—they preferred a military response—but Nasser continued to call for "sustained political pressure against Israel and the consolidation of Arab unity."[37] When little came of these efforts, he turned to the prospect of

[33] Interview with senior Egyptian nuclear expert, Cairo (March 2004).

[34] For an extensive discussion of the Kennedy-Nasser correspondence, see Heikal, *The Cairo Documents*, pp. 187–224. On Kennedy's requests that Egypt remain non-nuclear, see ibid., p. 208 and Ya'ir Evron, "A Nuclear Balance of Deterrence in the Middle East", *New Outlook* (July-August 1975), p. 27.

[35] Interview, former senior Egyptian science official, Cairo, March 2004.

[36] For example, at the Arab Foreign Ministers meeting in Baghdad in February 1961, "Arab representatives at the United Nations were to urge the United Nations to entrust the International Atomic Agency to investigate the Israeli situation." Evron, "A Nuclear Balance of Deterrence," p. 21. See also Flapan, Simha Flapan, "Nuclear Power in the Middle East", *New Outlook* (July 1974), pp. 39–40.

[37] Roger F. Pajak, "Nuclear Status and Policies of the Middle East Countries", *International Affairs*, Vol. 59, No. 4 (Autumn 1983): p. 591.

using the emerging nuclear non-proliferation regime as another way to assert Egyptian leadership via the Israeli nuclear issue. Egypt had already gained prominence due to its early and active involvement in disarmament issues.[38] The Soviets suggested that Egyptian signature of the NPT would embarrass Israel. Nasser's signature of the NPT in 1968 was not likely motivated by anti-nuclear or pacifist sentiments—rather, it was a strategic choice to use diplomatic tools to isolate Israel. Thus, Nasser's orientation toward the non-proliferation regime is best characterized as social conformity, given that he had not become persuaded of the value of non-proliferation, but rather was going along with the norm to maximize social benefits.

With Nasser's death in 1970, Anwar Sadat took over and kept the nuclear program frozen—his focus was on regaining the Sinai. Then Sadat made the strategic decision to close the nuclear weapons program permanently, due to his new approach to Egypt's future. He turned Egyptian foreign policy on its head, by seeking peace instead of war with Israel, an alliance with the US instead of the Soviets, an Egypt-first approach instead of Pan-Arabism, and economic development instead of socialism. Sadat's choices were counter-intuitive, placing him "at odds with the dominant Arab thinking, and certainly with Syria, his compatriot in the 1973 War, who continued to sponsor a rejectionist stance."[39]

Sadat's new vision led to a reframing of who Egypt's friends and enemies were, which also led to a rethinking of the Egyptian security dilemma. This opened the door to Egyptian polices that no longer valued nuclear weapons. Indeed, Sadat was prepared to offer the nuclear issue as an incentive to the United States. "Sadat knew that the U.S. would not come to Egypt without incentives. The nuclear issue was one of those issues we were ready to abandon. It was a tool, something we could give to the U.S. as a present, even though Sadat knew that the nuclear program was already closed and it wasn't really

[38] Egypt was a member of the Committee on Disarmament, created in 1961, which concluded the Partial Test Ban Treaty in 1963. Egypt also participated in the Committee on Disarmament from its establishment in 1965. See Mostafa Elwi Saif, "Egypt: A Non-Nuclear Proactivist In a Volatile Region," paper presented at the Peace Research Institute's Conference "The Domestic Roots of Proactivist Non-Nuclear Policy: A New Approach to Non-Proliferation," Bellagio, Italy (September 29–October 3, 1997), pp. 11–12.

[39] Michael Barnett, *Confronting the Costs of War: Military Power, State and Society in Egypt and Israel* (Princeton, NJ: Princeton University Press, 1992), p. 128.

much of a present."[40] This quotation illustrates the dramatic change in Egyptian security calculations. For a country facing a nuclear-armed adversary against which it had lost four wars, and knowing of secret nuclear weapons programs of three other neighbors, a nuclear capacity sounds like the perfect present. But with Sadat's new goals, military conflict with Israel was no longer planned. Instead, peace with Israel was the priority—and such a priority meant that nuclear weapons were no longer needed.

For this reason, the best explanation for Egypt's nuclear restraint is persuasion. Sadat reconceptualized Egyptian security, including what constituted a legitimate way to deal with an enemy, and this led to a devaluation of nuclear weapons. In this case, the most important factor is elite thinking. Sadat's truly radical changes in numerous facets of Egyptian policy are what led to complete and final closure of the nuclear weapons program. Such an explanation may seem ultimately unsatisfying to political scientists, who prefer generalizable lessons that extend across cases. However, it is important to note that without the international norm against nuclear acquisition, embedded in the NPT, Sadat would not have been required to discard nuclear weapons. Indeed, if instead nuclear weapons had become the hallmark of advanced militaries, Sadat likely would have pursued them vigorously. But the international norm against nuclear weapons shaped Sadat's choices, making "nuclear weapons" a costly choice that was not even desired.

After Sadat's assassination in 1981, Hosni Mubarak took power in Egypt. In terms of nuclear policy, his presidency was marked by two trends: attempted acquisition of nuclear power, and consistent use of institutions to pressure Israel to end its nuclear weapons program. With the first trend, Cairo made serious attempts to acquire civilian nuclear power reactors to meet Egypt's rapidly growing energy needs. It is interesting that as Israel's muscular nuclear program came to light, Egypt responded by adhering more firmly to the NPT and being quite clear that all nuclear power plants would be placed under the inspections of the International Atomic Energy Agency (IAEA). While Egypt expressed great interest in civilian nuclear power, it has yet to build a power plant, due to financial difficulties and then the abandonment of attempts following Chernobyl.[41]

[40] Interview with former senior Egyptian military officer, Cairo (March 2004). On Sadat's assurances to the US that Egypt had closed its nuclear option, see "Nuclear power for Cairo," *New York Times* (June 17, 1974), p. 30.

[41] For the most comprehensive review of nuclear power planning in Egypt, see Walsh, "Bombs Unbuilt."

In September 2006, Mubarak's eldest son and presumed heir, Gamal Mubarak, proposed a revival of Egypt's nuclear power plans, calling for plants to be operational in ten years.[42] The government began recruiting Egyptian scientists from overseas to work on the project,[43] and Egypt engaged in a flurry of preliminary negotiations with a number of countries, including the US, Russia, and China, for the plants.[44] The initial goal was to build three power plants by 2020, to provide 1,800 megawatts of energy.[45] However, as of 2011, only one contract has been awarded. In 2009, Cairo signed a ten-year, US$200-million contract with Worley Parsons: "the Australian company will be responsible for selecting locations, choosing reactor technology, ensuring quality control, training personnel to run the plant and providing miscellaneous technical services."[46] Planning has run into challenges: business interests argued that the most likely site for the first nuclear power plant (Al-Dabaa, on the Mediterranean coast) would be better used for tourism than for nuclear facilities.[47] In August 2010, Mubarak sided against the business interests and gave approval for construction in Al-Dabaa.[48] Egypt's nuclear power plans received the blessing of the United States and the IAEA. In light of Egypt's progression to site selection, IAEA head Yukiya Amano proposed sending a mission to Cairo for technical assistance, saying, "The IAEA is very happy to cooperate with Egypt in its project of introducing nuclear power."[49]

However, signals from Egypt do elicit some concern about the possible direction of the program. In 2007, Cairo stated that it intended to leave the door for uranium enrichment open, and that it would not sign the Additional

[42] "Egyptian president's son proposes developing nuclear energy," *Associated Press* (September 19, 2006).

[43] Ibid.

[44] Derek Sands, "Egypt looks East for nuke power," *United Press International* (November 13, 2006).

[45] "Egypt to start building nuclear power plants soon, minister says," *Associated Press* (September 24, 2006).

[46] Michael Janda, "Australian firm to design Egypt's nuclear future," *ABC News* (June 19, 2009).

[47] Gamal Essam El-Din, "A nuclear falling out," *Al-Ahram Weekly*, Issue 963 (Cairo: September 2009).

[48] "Egypt: 1st nuclear plant site announced," *Associated Press* (August 25, 2010) at http://www.jpost.com/MiddleEast/Article.aspx?id=185934.

[49] "U.N. watchdog backs Egypt nuclear power plant plans," *Reuters* (June 22, 2010) at http://www.reuters.com/article/idUSLDE65L0F320100622.

Protocol.[50] The refusal to sign that Protocol was reiterated in 2008.[51] Given that Egypt already has highly trained nuclear scientists and a large nuclear complex, the ability to create a breakout capability is somewhat of a concern:

> It has two research reactors, including a high-neutron flux Multi-Purpose Reactor (MPR) located at the Inshas research center. The 22 thermal megawatt (MWth) MPR could produce 6.6 kilos of plutonium a year assuming 300 days of operation. Egypt also has since 1998 two fuel-making installations, a laboratory (Semi-Pilot Fuel Laboratory) and a full-fledged plant (Fuel Manufacturing Pilot Plant) which fabricates fuel for the MPR reactor. It also has significant uranium reserves. As is now known, Egypt has made experiments that could be useful to a military program, before and after its safeguards agreement came in force (1982). These involved the production of uranium metal, irradiation of natural uranium targets, and chemical dissolution of fuel elements. The Atomic Energy Agency (AEA) has hot cells which could be used for experimental plutonium separation, located at a dedicated facility (Hot Laboratory and Waste Management Center).[52]

Thus, as Tertrais notes, Egypt is one of the few states in the Middle East capable of creating a "full-fledged, autonomous nuclear program."[53]

The second trend during this period was Egypt's use of the NPT and associated international instruments to put pressure on Israel to end its nuclear weapons program. From the 1970s, Egypt made dozens of attempts to blunt Israel's nuclear advantage through the use of UN and other international fora. From 1974 onward, Egypt sponsored a resolution in the General Assembly of the United Nations calling for a nuclear-weapon-free zone in the Middle East. In 1990, Mubarak proposed a weapons of mass destruction (WMD) free zone in the Middle East. Egypt brought the Working Group on Arms Control and Regional Security (ACRS) to a halt in 1995 by insisting that any progress must be linked to Israel's nuclear disarmament. With the 1995 NPT Review

[50] The Additional Protocol was enacted by the IAEA in the 1990s in order to "strengthen the effectiveness and improve the efficiency of the safeguards system". See "IAEA Safeguards Overview: Comprehensive Safeguards Agreements and Additional Protocols" at http://www.iaea.org/Publications/Factsheets/English/sg_overview.html.

[51] Leonard S. Spector and Benjamin Radford, "Algeria, Emirates Plan Nonproliferation-Friendly Nuclear Programs; Egypt Keeps Fuel Cycle Options Open, Rejects Expanded IAEA Monitoring," *WMD Insights* (June 2008) at http://www.wmdinsights.com/I25/I25_ME1_AlgeriaEmirates.htm.

[52] Bruno Tertrais, "The Middle East's Next Nuclear State," *Strategic Insights*, Vol. 8, No. 1 (January 2009).

[53] Ibid.

and Extension Conference, "Egypt hit the high pitch on nuclear non-proliferation," according to a former US diplomat. Egypt helped convince the other Arab countries to sign the NPT, in exchange for the Middle East resolution. With the failure to implement the 1995 Middle East resolution, Egypt played the role of spoiler in the 2005 NPT Review Conference, contributing in part to the conference's lack of a concluding consensus document. In the most recent 2010 compromise, Cairo held Washington hostage in Obama's quest for a meaningful consensus document. Not only was Cairo successful in negotiating for a conference on a nuclear weapons free zone (NWFZ) in the Middle East, but Egyptian diplomats refused to agree to a consensus document unless Israel was specifically named in the statement. "At the 11th hour, the U.S. team—headed by Ambassador Susan Burk—agreed not to block consensus on the entire document over this one point."[54] Some analysts have argued that Egypt exercised remarkable leadership at the conference and "played its hand brilliantly."[55]

What explains Egypt's continued faith in institutions to manage its security concerns? On the one hand, Sadat's redefinition of friends and enemies has become embedded in Egyptian elite's thinking. Israel was not seen as a serious security threat, although diplomats have been unhappy with the way Tel Aviv uses its undeclared nuclear weapons for leverage in the Palestinian situation. Rather, Iran's acquisition of nuclear weapons has been seen as much more serious, because Cairo has believed Iran would seek to become the regional hegemon. Others have noted that if Iran acquires nuclear weapons, then it becomes the natural counterpoint to Israel, displacing Egypt.[56]

Another reason for Egypt to continue its policy of nuclear restraint is its "sunk costs" in institutions. While Egypt gave up seeking the glory of the Arab bomb, it replaced it with the glory of leading the nuclear non-proliferation effort in the Middle East. A US diplomat said, "Some say that states pursue nuclear weapons for the glitter factor. For Egypt, the glitter factor was transferred over to using the UN and taking a leadership role diplomatically, espe-

[54] Elaine M. Grossman, "Egypt Plays Key Nonproliferation Role, But Keeps Nuclear Options Open," *Global Security Newswire* (June 10, 2010) at http://gsn.nti.org/gsn/nw_20100610_8351.php.

[55] Rebecca Johnson, quoted in Grossman, "Egypt Plays Key Nonproliferation Role".

[56] In June 2010, the Egyptian Ambassador to the United Nations made a veiled threat that Cairo would also pursue a military nuclear capability if Iran is allowed to succeed. Ibid.

cially with regards to the nuclear non-proliferation issue."[57] Although a former Egyptian Ambassador to the United States, Nabil Fahmy, has stated that the NPT does not work, one has to then question why Egypt continued to put its faith in the international nuclear non-proliferation regime to press Israel and now Iran. If Egyptian diplomats were genuinely convinced that the NPT was a failure and could not work, would one not expect that they would try something else? Instead, plans for nuclear power plants continue to linger, and Egyptian diplomats continue to shine at the United Nations.

These two reasons lead to the conclusion that persuasion is the best explanation for Egypt's continued nuclear restraint under Mubarak. Exactly what is Egypt persuaded of? One could argue that Cairo's rigid and shrill diplomacy regarding Israel's nuclear weapons makes it seem less likely that Egypt is persuaded about the genuine promise of nuclear non-proliferation and disarmament. Rather, it may seem like a return to Nasser's strategy: the NPT is a diplomatic tool with which to harass and pressure Israel. I would argue that this may be so, and in fact it is an important lesson for the future viability of the nuclear non-proliferation regime. Those who are persuaded may, over decades (or less), lose faith if institutions continually fail to live up to their promises. I argue that at least some members of Egypt's diplomatic corps remain genuinely persuaded as to the value of the NPT:

> One Western expert who has worked with the Egyptians noted, "Egypt has strongly associated itself with international law and international norms. There's a whole generation of Egyptian Foreign Ministry guys who were brought up and trained in the nuclear nonproliferation regime. They are not going to against the norms underlying the regime." Speaking of one of Egypt's most senior diplomats, a U.S. expert said, "Violating the NPT would undermine every personal value he holds. He spent a career supporting nuclear nonproliferation." Because the NPT took the issue of nuclear weapons out of the military's domain and into that of the Foreign Ministry, the beliefs and assumptions of these diplomats matter a great deal in Egypt's nuclear decision-making.[58]

Mubarak's caution allowed his diplomats to take the lead on the nuclear issue, and they have done so quite passionately. However, if institutions do not perform as expected, those who remain genuinely persuaded may become outnumbered by those who have lost faith and either advocate nuclear restraint out of conformity, or advocate outright rejection.

[57] Interview with US government official, Washington, DC (February 2004).
[58] Rublee, *Nonproliferation Norms*, p. 142.

Leadership transitions and nuclear futures in Egypt

With the ousting of Mubarak in February 2011, Egypt's nuclear future has been plunged into uncertainty, with possible outcomes ranging from a complete shutdown of all plans for nuclear power to a defiant withdrawal from the NPT and pursuit of a nuclear weapons program. Key to sorting through these possible futures is understanding who might lead Egypt and what their views are regarding the nuclear non-proliferation norm. The three most likely scenarios in post-Mubarak Egypt are a continued military junta, election of a secular leader such as Mohamed ElBaradei or Amr Moussa, and domination by Islamist groups such as the Muslim Brotherhood.

The military junta seems to be making progress toward a shift to free elections. However, the process could be prolonged in the hope of weakening support for reform and distracting the public with other issues.[59] If the military manage to stay in power, what type of nuclear policies might they pursue? Given that the military are strongly invested in the status quo, probably they would not seek any radical changes that could threaten the stability of the regime. The military elite is not motivated by religious or political agendas, but rather by the desire to maintain its privileged position within Egyptian society. For example, the military controls not only security and the defense industry, but production of many consumer goods:

> The military has built a highway from Cairo to the Red Sea; manufactures stoves and refrigerators for export; it even produces olive oil and bottled spring water. When riots broke out during bread shortages in March 2008, the army stepped in and distributed bread from its own bakeries. Military Factor 99 ... produces a variety of consumer goods—stainless steel pots and pans, fire extinguishers, scales, cutlery—in addition to its primary function of forging metal components for heavy ammunition.[60]

Ensuring that their institutional and industrial dominance in Egypt remains unthreatened will be a top priority for the military; in this case, there would be no pressing reason for dramatically altering course on nuclear issues. This is particularly the case for Field Marshal Mohamed Hussein Tantawi, the head of the Supreme Council of the Armed Forces, who has been described

[59] See, for example, "Egypt's military leadership is reacting too slowly" (interview with Mohamed El Baradei), *Der Spiegel* (April 19, 2011) at http://www.spiegel.de/international/world/0,1518,757786,00.html.

[60] Thanassis Cambanis, "Succession gives army a stiff test in Egypt," New York Times (September 11, 2010).

as change-resistant and focused on the status quo.[61] Indeed, a likely result could be a shelving of the plans for nuclear power due to financial constraints. Projected loss of tourism revenues and foreign direct investment from the political turmoil led the corporation Moody's to revise its outlook on Egypt to negative from stable.[62] However, the military are unlikely to ignore the nuclear issue completely and could potentially flirt with the prospect of a nuclear weapons program under certain circumstances. One of the leaders of the Supreme Council, Lt. Gen. Sami Hafez Enan, Chief of Staff of the Armed Forces, previously voiced concern over Iran's growing influence in the region.[63] If Iran manages to acquire nuclear weapons, there could be interest among the Egyptian military in matching Iran's capability. Additionally, domestic political pressure could lead the military to at least talk about the possibility of a nuclear weapons program. The Egyptian populace is quite unhappy with Israel's monopoly on nuclear weapons; in fact, a majority almost certainly support an Iranian nuclear weapons program, not to mention an Egyptian nuclear weapons program.[64]

Given this, the best characterization of the military toward the nuclear non-proliferation norm is that it is conforming. Egyptian generals are not particularly passionate about nuclear non-proliferation; the military's concern is retaining "its pre-eminent position in the nation's affairs," and the same is true of the military's current leader Field Marshal Tantawi.[65] Thus, persuasion would not characterize their adherence to the non-proliferation norm. Identification also seems unlikely, as the military do not seem to be trying to pattern their nuclear policy in relation to an important other. However, one could argue that while the military might naturally be interested in a nuclear weapons program, especially because of Iran's flirtations, they are more likely

[61] Yolande Knell, "Egypt after Mubarak: Mohamad Hussain Tantawi profile," *BBC News* (February 12, 2011) at http://www.bbc.co.uk/news/world-middle-east-12441512.

[62] "Moody's puts Egypt banks on negative outlook," *Reuters* (April 19, 2011) at http://www.reuters.com/article/2011/04/19/egypt-moodys-idUSLDE73I1KW20110419.

[63] Thom Shanker and Eric Schmitt, "Egypt's military leaders face power sharing test," *New York Times* (February 11, 2011).

[64] Mohammad Abdel Salam, "The Arab position on Iranian nuclear issues," *Al-Ahram Center Egyptian Commentary*, Issue 21 (Cairo: September 9, 2004) at http://www.ahram.org.eg/acpss/eng/ahram/2004/7/5/EGYP25.HTM.

[65] Thanassis Cambanis, "Succession gives army a stiff test in Egypt."

to remain non-nuclear because of the costs of going against the norm. Such a position is best understood as one of conformity.

More likely than not, however, the military will be pressured into holding elections under international supervision. If that happens, a secular leader such as Mohamed ElBaradei or Amr Moussa could win the presidency. What would the repercussions be on Egypt's nuclear policies? Former Egyptian Ambassador to the United States Nabil Fahmy argues that little would change:

> Neither Moussa nor ElBaradei are expected to dramatically change Egypt's foreign policy, and I don't believe that anyone who can get elected nationally would. Whomever is elected, however, will recalibrate it with a stronger national and proactive focus. Either Moussa or ElBaradei would retain a centrist approach but make it tougher and more constructive. Both are "achievers." They don't like remaining on the sidelines, or being negative, and neither constitutes a threat or would change Egypt's positions on WMD issues and the NPT.[66]

ElBaradei's leadership at the IAEA won him a Nobel Peace Prize, and his several years in the agency have led him to become a champion of both non-proliferation and disarmament. While he did not support Egypt's membership in the NPT initially, he advocates fully adhering to all treaty obligations. In contrast, Moussa believed in Egypt's NPT membership from the start.[67] One likely area of policy change is increased pressure on Israel to join the NPT, although Moussa is more likely to press for immediate changes whereas ElBaradei is more likely to link Israel's nuclear issue with a comprehensive peace settlement. In either case, domestic political pressure would likely have less effect on Moussa and ElBaradei than on a military government; both men have intellectual commitments to non-proliferation that would raise the bar for such action. However, if there was significant domestic pressure on the government to match Israel's nuclear program, both Moussa and ElBaradei could respond with much more strident positions on the right to a full nuclear fuel cycle, the voluntary nature of the Additional Protocol, and more.

In contrast to the military, both ElBaradei and Moussa are best characterized as persuaded. ElBaradei certainly would not be seen as "identifying" with the United States, given his public opposition to the Iraq war. Nor does con-

[66] James Martin Center for Nonproliferation Studies, "The Egyptian Revolution and the Future of WMD in the Middle East: Q&A With Nabil Fahmy" at http://cns.miis.edu/activities/110228_fahmy_egypt_qna.htm.

[67] Ibid.

formity describe ElBaradei, for that would imply that he would pursue a nuclear weapons option if major diplomatic costs could be avoided. Moussa argued for Egyptian membership in the NPT from the beginning, and supported Mubarak's use of diplomacy to counter Israel's nuclear program. In addition he would not be described as "identifying" with Washington, nor has he argued for nuclear hedging that one might expect from someone who was conforming due to diplomatic pressure. However, both men could be moved at least partially from their persuaded position if the promised 2012 Conference on NWFZ in the Middle East does not materialize or fails to move Israel's position.

The final likely scenario involves increasing Islamist domination of Egyptian politics, most likely through the Muslim Brotherhood. The influence of the formerly banned political party is evident. Just one month after Mubarak stepped down, the "young, educated secular activists were no longer the driving political force," with some suspecting that the military had "struck a deal" with the Muslim Brotherhood to blunt fundamental changes.[68] The early influence of the group is most evident in the March 2011 referendum on constitutional amendments. A member of the Muslim Brotherhood was appointed to the committee that created the amendments for the referendum, and the group was very effective at gathering votes for its preferred outcomes, using appeals to Islam.

> The amendments essentially call for speeding up the election process so that parliamentary contests can be held before September, followed soon by a presidential race. That expedited calendar is seen as giving an advantage to the Brotherhood and to the remnants of Mr. Mubarak's National Democratic Party, which have established national networks. With a record turnout, the vote was hailed as a success. But the "yes" campaign was based largely on a religious appeal: voters were warned that if they did not approve the amendments, Egypt would become a secular state.[69]

The Muslim Brotherhood initially promised that it would not put up a candidate for the presidential elections. However, in May 2011, Muslim Brotherhood leader Abdul Munim Abu al-Futuh declared he would run for president as an independent candidate.[70] The Muslim Brotherhood announ-

[68] Michael Slackman, "Islamist group is rising force in a new Egypt," *New York Times* (March 24, 2011).

[69] Ibid.

[70] "Muslim Brotherhood leader to run for Egyptian presidency," *Herald Sun* (Melbourne: May 11, 2011) at http://www.heraldsun.com.au/news/breaking-news/muslim-

ced it would not back any independent candidate who was a member of its movement, but should a leader of the group run, the perception of endorsement may be hard to overcome.

How would the Muslim Brotherhood influence Egypt's nuclear policies? Some, such as Nabil Fahmy, argue that the political party would respect existing agreements, unless Israel openly declared its nuclear weapons program.[71] However, the Muslim Brotherhood has been quite vocal in its support for an Egyptian nuclear deterrent. The group has publicly stated that an Iranian nuclear weapon will be an Islamic deterrent against Israel.[72] Indeed, it wants Egypt to develop nuclear energy—and nuclear weapons—to balance Israel. As the spokesperson for the Muslim Brotherhood parliamentary caucus said in July 2006, "We are ready to starve in order to own a nuclear weapon that will represent a real deterrent and will be decisive in the Arab-Israeli conflict."[73] Additionally, the Muslim Brotherhood has argued that Egypt "should have nuclear weapons as a matter of national right and prestige."[74] The group attempted to use Mubarak's support of non-proliferation to undermine his regime, saying that Egypt should balance Israel rather than placate it.[75] If the Muslim Brotherhood were to gain power in Egypt, its past stance on the issue indicates that it would likely work toward a military nuclear capability. The Brotherhood's view of the nuclear non-proliferation norm is clear: it rejects it.

Should Abu al-Futuh win the presidency, he would probably not deviate from the Muslim Brotherhood position on nuclear weapons. He has been described as "pragmatic" and has openly stated that Muslims and Coptic Christians are "partners" together in Egypt.[76] He also argued that it was the

---

brotherhood-leader-to-run-for-egyptian-presidency/story-e6frf7jx-1226054262936.

71 "The Egyptian Revolution."

72 Lynda Hurst, "Is Mideast on brink of an arms race?" *Toronto Star* (Toronto: January 27, 2007).

73 Sammy Salama and Khalid Hilal, "Egyptian Muslim Brotherhood Presses Government for Nuclear Weapons," *WMD Insights* (November 2006) at http://www.wmdinsights.com/I10/I10_ME3_EgyptianMuslim.htm.

74 Hurst, "Is Mideast on brink of an arms race?"

75 Salama and Hilal, "Egyptian Muslim Brotherhood Presses Government for Nuclear Weapons."

76 Amr Hamzawy, Marina Ottaway and Nathan J. Brown, "What Islamists Need to be Clear About: The Case of the Egyptian Muslim Brotherhood," *Carnegie Endowment for International Peace Policy Outlook* (February 2007).

duty of Muslims to fight Americans in Iraq and Israelis in Palestine. In an interview on Al-Jazeera TV (Qatar), Abu al-Futuh said, "I personally, as all members of the Muslim Brotherhood in Egypt and around the world, wish to carry weapons against the American occupation in Iraq, just as we wish to carry weapons against the Zionist enemy in Palestine. When a Muslim land is occupied, jihad becomes an individual duty for every man and woman, boy and girl."[77] Abu al-Futuh's first foreign policy statement as a presidential candidate also indicates that he seek to reverse Mubarak's diplomacy. In the Egyptian daily *Al-Shourouq*, he argued that all Egyptians wanted "no normalization with Israel, and want an approach opposite to that of the previous regime, that acted according to external dictates."[78]

Leadership, transitions and nuclear decision making in Libya

Libya's nuclear history[79]

Even more so than Egypt, the history of Libya's quest for nuclear weapons is dominated by the importance of individual actors. After Muammar Qadhafi took power in Libya in 1969, he began his drive for a military nuclear capability. At first Qadhafi concentrated on purchasing nuclear weapons or components, but after a decade of failed attempts, he switched his focus to developing an indigenous capacity. A number of efforts were made to acquire nuclear technology, train Libyan students in nuclear science, and even recruit former

[77] "Dr. Abd Al-Munim Abu Al-Futuh, an Egyptian Muslim Brotherhood Leader, Volunteers to Carry Out Attacks Against American Forces in Iraq," *Al-Jazeera TV* (Qatar: April 25, 2004), translated by MEMRI at http://www.thememriblog.org/blog_personal/en/31129.htm.

[78] *Al-Shourouq* (Egypt: May 12, 2011), reported in "Muslim Brotherhood Official To Run As Independent For Egyptian Presidency," translated by MEMRI at http://www.thememriblog.org/blog_personal/en/37468.htm.

[79] This section on Libyan nuclear history is drawn from research for my book, *Nonproliferation Norms*, chapter 5. Other important works on Libyan nuclear issues include Målfrid Braut-Hegghammer, "Libya's Nuclear Turnaround: Perspectives from Tripoli," *Middle East Journal*, Vol. 62, No. 1 (2008); Ray Takeyh, "The Rogue Who Came in from the Cold," *Foreign Affairs*, Vol. 80, No. 3 (May-June 2001); Wyn Q. Bowen, *Libya and Nuclear Proliferation: Stepping Back from the Brink*, Adelphi Papers, No. 380 (London: International Institute for Strategic Studies, May 2006); Solingen, *Nuclear Logics*; and Bruce W. Jentleson and Christopher A. Whytock, "Who 'Won' Libya? The Force-Diplomacy Debate and Its Implications for Theory and Policy," *International Security*, Vol. 30, No. 3 (Winter 2005–6).

Soviet scientists to come to Libya. However, tightened export controls and moves by the US made progress difficult. By the mid-1990s, Qadhafi's foreign policy was being pulled in two directions. On the one hand, Tripoli had lost its major benefactor with the collapse of the Soviet Union, and economic sanctions were weighing heavily on the economy. Reformers within Libya were advocating economic openness, a less bellicose foreign policy, and domestic political reforms. Chief among them was Qadhafi's second eldest son, Saif al-Islam al-Qadhafi, who pushed especially hard for media freedoms and humanitarian causes. On the other hand, conservatives feared a turn away from the regime's radicalism would result in a loss of domestic legitimacy; they also believed they could lose privilege in a reformed Libya. The result was a schizophrenic approach to security policy, veering from overtures to the British in hopes of reconciliation to continued imports of sensitive nuclear technology through the black market.

The conflict between reformers and revolutionaries continued until 1998, when Qadhafi signaled clear support for reform. Takeyh notes,

> A series of articles in the official daily Al-Jamahiriya began to criticize the intransigence of the hardliners and their inability to recognize prevailing global realities. The Revolutionary Committees—informed groups of zealots ... that served as the hardliners' power base and had dominated Libyan politics since the late 1970s—were purged and relegated to the margins of society.[80]

A year later, Qadhafi surrendered the two suspects in the Lockerbie bombing attack and twice approached the United States offering to give up his WMD programs.

Washington did not take Tripoli's offers seriously, but Qadhafi's cooperation with the British led to increasing pressure on the United States to drop its objection to the removal of United Nations sanctions. As a result, "Clinton decided to begin negotiations with Libya to see if the United States could 'graduate' the rogue state into the international community."[81] However, the United States was focused on Libya's terrorist connections and put the issue of WMD on the backburner. Salama notes, "It was assessed that Libya's modest chemical weapons arsenal and infant nuclear program were not an imminent threat, and as a result, there was no urgency driving the US to accept Libya's offer to surrender its WMD."[82] Negotiations stalled under the Bush

[80] Ray Takeyh, "The Rogue Who Came in from the Cold," *Foreign Affairs*, Vol. 80, No. 3 (May-June 2001): p. 66.

[81] Rublee, *Nonproliferation Norms*, p. 155.

[82] Sammy Salama, "Was Libyan WMD Disarmament a Significant Success for Non-

administration; in early 2003, Libyan diplomats reiterated their desire to talk. In October 2003, Western agents interdicted a German ship bound for Libya with thousands of centrifuge parts. Finally, Washington was ready to talk, and in December 2003, Qadhafi announced he would give up all materials related to the nuclear weapons program. Since that time, Tripoli has cooperated fully with the West and the IAEA in dismantling its program.[83]

The argument that Libya gave up its nuclear weapons because of threat of invasion does not accord with the historical timeline, given that Qadhafi began offering to give them up in 1999. Others argue that international sanctions forced Qadhafi's hand. Clearly, sanctions put pressure on Tripoli; however, again, the historical timeline shows that sanctions were not the critical factor:

Action	*Date*
UN sanctions suspended	April 1999
Qadhafi offers to give up WMD programs	May 1999, October 1999
UN sanctions permanently lifted	September 2003
Qadhafi gives up WMD programs	December 2003

Thus, offers to give up WMD could not have been motivated by a desire to see the sanctions lifted, because that had already taken place. In addition, as one scholar has argued, sanctions against Libya increased its isolation, which made it difficult for Tripoli to chart a more cooperative path, and also radicalized its nuclear decision making.[84] Thus, while sanctions motivated reform-minded elements to push for reform, they also invigorated hardliners in their pressure for continued isolation.

---

proliferation?", *Issue Brief* (Monterey Institute of International Studies, September 2004) at http://www.nti.org/e_research/e3_56a.html.

[83] The odd exception to this cooperation took place in November 2009, when Libya refused to turn over the last 5.2 kilograms of weapons-grade uranium to the Russians. After a phone call from US Secretary of State Hillary Clinton to Libyan Foreign Minister Musa Kusa, the Libyans allowed the uranium to leave the country. For more information, see Max Fisher, "A Nuclear Standoff With Libya," *The Atlantic* (Washington, DC: November 2010).

[84] Målfrid Braut-Hegghammer, "Libya's Nuclear Turnaround: Perspectives from Tripoli," *Middle East Journal*, Vol. 62, No. 1 (2008): pp. 68–9. See also Michael Collins Dunn, "MEJ Author Målfrid Braut-Hegghammer on Libya's Nuclear Rollback," *MEI Bulletin*, Vol. 59, No. 1 (March 2008): pp. 8–9.

Critical to Libya's turnaround, not just on the issue of WMD but also on the issue of engagement with the international community more generally, was Saif al-Islam, Libya's "main referent of Libyan reformism."[85] As Pargeter notes,

> Those within the reformist camp may have had more progressive ideas about the future of Libya than other traditionalists within the regime who felt that their own privilege was tightly linked to maintaining the old socialist style Jamahiriyah. However, these reformists were still essentially part of the older generation and many had been well entrenched within the regime since the revolution of 1969. It was the coming to age of Qadhafi's eldest son by his second wife, Saif al-Islam, that brought new blood into the equation and that really added impetus to the calls for internal reform.[86]

Saif al-Islam led the initiative for Libya to renounce its nuclear program, against the opposition of hardliners.[87] Indeed, it is important to note that it was not just reformers in general, but Saif al-Islam in particular, who "coaxed" his father into giving up the pursuit of nuclear weapons.[88] Saif al-Islam likely had particular leverage with his father, given that he had been seen as a favorite to succeed Qadhafi.

Libya's current nuclear status

Tripoli cooperated fully in dismantling its nuclear weapons programs; both the British and the IAEA praised Libya's openness and progress. The one exception to this cooperation occurred in late 2009, when the last of Libya's weapons-grade (highly enriched) uranium (HEU) was delayed in its shipment out of the country to Russia. According to US State Department documents, the materials were ready to be loaded onto a plane specially equipped to fly nuclear materials when Libyan officials halted the shipment.[89] Within a few days, Saif al-Islam contacted the US Embassy in Tripoli to demand concessions if the uranium were to be removed, including "military equipment

85 Arturo Varvelli, "Italy and Libya: Renewing a Special Relationship," *The International Spectator*, Vol. 45, No. 3 (2010): pp. 124.

86 Alison Parteger, "Libya: Reforming the Impossible?" *Review of African Political Economy*, Vol. 33, No. 108 (2006): p. 222.

87 "US embassy cables: Gaddafi's son links uranium crisis to father's 'humiliation' in New York," *The Guardian* (London: December 3, 2010) at http://www.guardian.co.uk/world/us-embassy-cables-documents/237232.

88 Vivienne Walt, "Can Gaddafi's Son Reform Libya?", *Time* (New York: April 5, 2010) at http://www.time.com/time/magazine/article/0,9171,1975241,00.html.

89 See Fisher, "A Nuclear Standoff with Libya" and "US embassy cables," *The Guardian*, for details.

(non-lethal and lethal weapons), an update on what was being done with Libya's centrifuges, movement on the Regional Nuclear Medicine Center, and financial assistance for the chemical weapons destruction program, including construction of the destruction facility."[90] In addition, Tripoli wanted an affirmation of the relationship between the two countries, in particular because Qadhafi had felt humiliated during his visit to the United Nations in New York, when he was not permitted to visit Ground Zero. After a meeting with the US Ambassador and a call from Secretary of State Hillary Clinton, Tripoli allowed the HEU to be flown to Russia.

While this incident could be seen as evidence of Saif al-Islam's hypocrisy on non-proliferation, more likely he was playing the role of intermediary. Conservative forces in Libya were unhappy with the dismantling of the WMD programs, and a lack of progress on numerous fronts with Washington led some to conclude that what Libya thought was a true turnaround in its relationship with the United States was merely "a public relations coup" for the Bush administration.[90] Given that this was the last of the HEU—and thus, some might consider, their last piece of leverage over the United States—such tactics are unsurprising. By leading the way on negotiations, Saif al-Islam could both smooth over the relationship with Washington (important if reforms in Libya would continue) and claim drawing a hard line with the United States to bolster his own domestic legitimacy.

Apart from this incident, Libya's nuclear policies have been uncontroversial. Its capabilities before the events of 2011 were not insignificant:

> Libya's nuclear program has been reduced to its civilian research component. However, Tripoli retains significant infrastructure and expertise, which would be useful if the country wanted launch a military program again. Its IRT-1 has been operational since 1981. It is a high neutron flux reactor, but one whose plutonium production remains limited because of its power (10 MWth). The Tajura Nuclear Research Center (TNRC) also comprises a number of experimental facilities: a 100 W neutron generator, a TMA-4A tokamak, and hot cells. Numerous undeclared experiments of uranium targets irradiation were conducted at Tajura between 1984 and 1990, with the production of small quantities of plutonium. ... Tripoli is keen on maintaining its expertise.[92]

In addition, Tripoli announced a nuclear power program, signing agreements with France, Argentina, the Ukraine, Russia and Canada for civilian

[90] Ibid.

[91] Ibid.

[92] Tertrais, "The Middle East's Next Nuclear State."

power reactors (for both energy and desalination).[93] The Libyan Atomic Energy Institute created a special panel to draft a law on nuclear power, to ensure the appropriate regulatory and legal frameworks were in place to support a civilian nuclear complex. In addition, another commission was charged with investigating sites for nuclear power plants.[94] Most recently, Libya signed a declaration of intent with France to form a strategic partnership, including bilateral talks between the two countries' nuclear agencies, the signature of a nuclear accord between Tripoli and Paris, and the construction of at least one nuclear power plant.[95]

Leadership transitions and nuclear futures in Libya

What does Libya's nuclear future entail? With the civil war, Libya's nuclear prospects have been in upheaval. As with Egypt, Libya's nuclear path will depend greatly on who leads the country in the post-Qadhafi era. At the time of this writing, in late summer 2011, the bloody civil war was not over, with central authority far from being established following Qadhafi's overthrow. Nevertheless, despite highly fluid and rapidly changing developments, a number of projections can be offered concerning the incoming regime's possible nuclear policy options.

Regardless of the international posture and priorities of Libya's post-Qadhafi leaders, financial constraints, especially in the wake of a devastating civil war, are likely to play a determining role in the available range of choices in the nuclear arena. Especially if the country adopts democratic institutions, elite focus will be on economic development and meeting the basic needs of people, who have been neglected by Qadhafi's policies. An expensive program such as nuclear power is unlikely to be high on anyone's list. However, over the medium term, a more aggressive nuclear program, perhaps with military dimensions, might still be a possibility. While the anti-Qadhafi opposition is not markedly religious, one of the more prominent rebel leaders, Abdel-Hakim al-Hasidi, has indicated links with al-Qaeda. Al-Hasidi fought against the Americans in Afghanistan, recruited Libyans to fight against the Americans in

93 "Libya moving forward with nuclear power plans," *World Nuclear News* (January 8, 2010) at http://www.world-nuclear-news.org/NP-Libya_moving_forward_with_nuclear_power_plans-0801107.html.

94 Ibid.

95 "Libya, France sign declaration of intent," *The Tripoli Post* (Tripoli: October 24, 2010).

Iraq, and is a member of the Libyan Islamic Fighting Group, a group that has had a cooperative relationship with al-Qaeda. Al-Hasidi said, "his fighters 'are patriots and good Muslims, not terrorists,' but added that the 'members of al-Qaeda are also good Muslims and are fighting against the invader.'"[96]

If al-Hasidi and his comrades eventually consolidated power in the post-Qadhafi era, they would likely have a similar perspective toward nuclear weapons as the Egyptian Muslim Brotherhood. Reforming Libya domestically to become more compliant with *shari'a* law would likely be higher on their list than pursuing a military nuclear capability, but the latter is not out of the question. This is especially the case because Libya lacks prominent secular intellectuals that Egypt has—with the exception of Saif al-Islam al-Qadhafi and his circle of old regime reformers, who will have little power to shape public debate.

Nuclear politics and the importance of individuals

In both Egypt and Libya, nuclear histories and nuclear futures clearly paint the importance of individuals. Nasser's ambivalence toward nuclear weapons meant that his drive for military nuclear capability faltered, when under another, more committed leader it might have flourished. Sadat completely rewrote Egypt's foreign policy, going to visionary lengths that simple economic reform did not dictate. In his rethinking of whose Egypt's friends and enemies were, he also rethought the value of nuclear weapons, deciding to permanently close the program. This closure stayed firm under Mubarak, who was happy to let the passionate diplomatic corps keep Egypt on a non-nuclear path. Looking to Egypt's nuclear future also shines the spotlight on why leadership transitions matter. The military elite is focused on maintaining its privileged positions in society. While both Iran's nuclear program and domestic pressure might tempt generals to think about their own military nuclear capability, the high diplomatic costs of doing so mean that Field Marshal Tantawi is likely to conform and continue Mubarak's adherence to the nuclear non-proliferation norm. Both ElBaradei and Moussa have staked careers on supporting the nuclear non-proliferation regime; if either were elected, Egypt would strongly engage in international nuclear diplomacy and it would be close to unthinkable that either would approve of a nuclear weapons program. On the other hand, should the Muslim Brotherhood, perhaps through the

96 Praveen Swami, Nick Squires and Duncan Gardham, "Libyan rebel commander admits his fighters have al-Qaeda links," *The Telegraph* (London: March 25, 2011).

presidency of Abdul Munim Abu al-Futuh, gain control in a post-Mubarak Egypt, one could expect the opposite: Egypt might withdraw from the NPT and actively pursue nuclear weapons. In other words, the choice of leadership matters a great deal to nuclear outcomes.

In Libya, Qadhafi sought nuclear weapons in his quixotic quest to become the "bad boy" of the Middle East. However, the influence of his son Saif al-Islam convinced him to rethink Libya's future, among other things, by renouncing the nuclear weapons program. As Qadhafi's regime crumbles, Islamist rebels such as Abdel-Hakim al-Hasidi would likely reject the non-proliferation norm, although their ability to pursue a nuclear weapons program would be severely hampered by the financial chaos that the aftermath of revolution will bring. Again, leadership transitions have clear and direct impacts on nuclear policy.

To what extent do other approaches explain Egyptian and Libyan nuclear politics? First, it is important to stress that this chapter is not arguing that individuals are the only significant variable. Elsewhere, I have argued that for both Egypt and Libya, the international non-proliferation norm shaped both individual and domestic debates about what success and security look like for those countries.[97] Nevertheless, this does not undermine the importance of understanding individual leaders' orientations regarding the nuclear non-proliferation norm. Both political economy and domestic coalitions are also clearly important to nuclear choices, as evidenced by both the nuclear histories and the nuclear futures of the two countries spelled out in this chapter. Etel Solingen's insight (in this volume) that internationalizing coalitions generally do not support military capability, while inward-looking coalitions do, informs analysis of both Egypt and Libya. However, examining elite orientations to the nuclear non-proliferation norm adds to, and sometimes challenges, this wisdom.

For example, both those who are persuaded and those who identify fit within Solingen's internationalizing type of coalition. However, should the nuclear non-proliferation norm erode significantly, someone who identifies may wander from nuclear restraint, because the commitment is not to the norm, but to an important "other." Should the US-India nuclear deal end up diluting nuclear export controls, over the long term, it may no longer be disadvantageous for business to seek military nuclear capability—in fact, it may become a way to increase exports. For internationalizers who "identify," this

[97] Rublee, *Nonproliferation Norms*, chapters 4–5.

could lead to erosion of their support for nuclear restraint. However, for an internationalizing elite that is persuaded, the value of nuclear weapons will not have changed. The commitment to security and security through peaceful means, rather than weapons, will not be as easily eroded. Thus, understanding individual orientations toward the nuclear non-proliferation norm offers finer gradients of the internationalizing type of coalition, with important implications for policy.

Table 3.1 Leadership Transitions and Nuclear Proliferation Outcomes

Coalition Type	*Internationalizing Coalitions*		*Inward-looking Coalitions*	
Nuclear position	*Persuaded*	*Identifying*	*Conforming*	*Rejection*
Application to Egypt	ElBaradei, Moussa	(Gamal Mubarak)	Mohamed Hussein Tantawi (Military elite)	Abdul Munim Abu al-Futuh (Muslim Brotherhood)
Application to Libya	(Saif al-Islam al-Qadhafi)		(Mutassim Qadhafi)	Abdel-Hakim al-Hasidi (rebel leader)
Proliferation Outcome under strong NPT	No proliferation	No proliferation	No proliferation	Attempted proliferation
Proliferation Outcome under weak/decaying NPT	No proliferation	Possible proliferation	Likely proliferation	Attempted proliferation

This chapter's framework also helps explain the anomaly of inward-looking coalitions that refrain from nuclear weapons. The Egyptian military controls not only security and the defense industry, but also many consumer goods, as noted earlier.[98] The military want to ensure that their institutional and industrial dominance in Egypt will not be threatened by excessive internationalization, and yet they have accepted Egypt's non-nuclear stance for decades. Why should an autocratic institution with inward-looking economic policies favor the opposite of what might be expected? The answer may lie in again understanding orientations toward the nuclear non-proliferation norm. Inward-

[98] Thanassis Cambanis, "Succession gives army a stiff test in Egypt."

looking coalitions may reject the norm, but they may also exercise conformity. In this case, if the perceived costs of pursuing nuclear weapons outweigh the benefits, the group may decide in favor of nuclear restraint—at least until the cost-benefit calculation changes.

In addition to helping us better understand the impact of domestic coalitions on nuclear policy, examining individual orientations toward nuclear weapons can introduce new policy prescriptions. In states dominated by personalities, finding ways to influence the leadership may be the most effective way to drive substantive and significant changes in policy.

4

The Role of Nuclear Energy in Gulf States' Economic Development

Giacomo Luciani

The world faces an energy and environmental challenge whose seriousness appears only to be increasing. For a time, emphasis in the international debate has been on greenhouse gas (GHG) emissions and the danger of global warming beyond the accepted maximum level of 2°C. The preoccupation for climate change has profoundly influenced our vision of energy futures, emphasizing the need to reduce dependence on fossil fuels and achieve transition toward new sources. At the same time, the rapid increase in energy demand from the emerging economies is outpacing the growth of supplies, including supplies of traditional fossil fuels, pointing to a future of increasing energy costs.

Although the European Union has officially set a goal of reducing emissions by 85–90 percent in 2050 relative to 1990, it is not at all clear that sufficient political will exists in the major economies to implement the policies that would be needed to maintain the concentration of CO_2 in the atmosphere within the limit of 450 p.p.m. (parts per million). Pledges that countries announced at the Copenhagen conference are not sufficient to reach this objective, and the political tide is turning against environmental concerns in several key countries, the United States first and foremost.

Reluctance to curb the consumption of fossil fuels contributes to preoccupations about their availability. Although resources are vast, the investment needed to bring them to the market—be it coal, oil or gas—is lagging, and supply is expected to become increasingly tight.

Energy demand in the context of GCC economic development

This global context is the unavoidable backdrop to our discussion of the energy future of the Gulf Cooperation Council (GCC) countries. Although their process of economic development and diversification has been neither uniform nor without unresolved issues, nevertheless their will to develop is not to be doubted. This should be the underlying premise of all discussion of the region's energy future.

Notwithstanding all difficulties inherent in the region's physical and climatic characteristics, the alternative of perpetuating the condition of rentiers by simply transforming hydrocarbon resources into financial assets globally invested is not accepted. The GCC countries do not aspire to retire in peace after their oil and gas resources are exhausted; they aim instead at developing diversified economies, which will support their growing populations and let them be counted among the advanced countries of the world.

Diversification in industry

The objective of economic diversification has been pursued since the early 1970s with a strategy based on openness to international trade and cooperation and tight integration in the global economy. The GCC countries have differentiated themselves from numerous other developing countries in that they never attempted to "do it alone," but always strove to have access to the best technology and know-how, and exposed their nascent industries to global competition with minimal protection.

What diversification has taken place has been achieved on the basis of the region's comparative advantages—the few that existed to begin with and those that were acquired through intelligent and bold investment. Although one frequently hears accusations that local industry is "subsidized"—hence not viable in the long run—it is increasingly being accepted that GCC companies in a number of key sectors are here to stay.

There are a number of sectors where GCC companies have been able to emerge as global competitive players:

- First in the list are the oil refining and petrochemical industries, where oil or gas are the essential feedstock. The GCC countries created considerable refining capacity in the 1970s and early 1980s, but later stopped expanding for a decade and a half. Interest in revamping existing refineries and establishing greenfield projects has emerged again in a big way in the past decade. A number of projects are underway, substantially increasing the region's total refining capacity.[1]

In contrast, investment in petrochemicals continuously grew since the 1970s, primarily in Saudi Arabia and Qatar, but increasingly so also in Abu Dhabi and to a lesser extent Oman and Kuwait. The initial round of petrochemical projects was based on cracking ethane and in some cases liquid petroleum gas (LPG), but the latest round includes primarily projects that are based on cracking naphtha and integrated with refineries.[2] In parallel, the industry has gained in complexity, moving away from production of simple plastics and basic building blocks into increasingly complex components.

Petrochemicals are a hugely important and diversified industry, and a valid form of economic diversification, because the region will have enough hydrocarbons to support a petrochemical industry for centuries rather than decades. Demand for petrochemical products is bound to remain very dynamic, as we live in the age of plastic, and there is no sign that this might change.

- A second line of industry in which the GCC countries have been highly successful is aluminum smelting.[3] The number and capacity of smelters in the region are growing rapidly, with the original projects in Bahrain and Dubai expanding and new projects having come on stream or being launched in Qatar, Abu Dhabi and Saudi Arabia.

[1] I have discussed the development and prospects of the GCC refining and petrochemical industry in a paper for the Al Jisr project of the Gulf Research Center, to be published in Giacomo Luciani (ed.), *Resource Blessed: the Diversification of the GCC Economies* (Dubai: Gulf Research Center, 2011).

[2] A review of investment projects underway can be found in the Annual Report 2010 of the Gulf Petrochemicals and Chemicals Association (GPCA), which is the representative body of petrochemical producers in the region. The GPCA estimates that the share of the Gulf producers in global petrochemicals production will increase from 15 percent in 2010 to 20 percent in 2015.

[3] Prospects for the aluminum industry in the GCC are discussed in Halvor Kvande, "The GCC Aluminium Industry—The Development of Aluminium Production Capacity in the Middle East" in Giacomo Luciani (ed.), *Resource Blessed*.

- Further lines of successful diversification have been the manufacture of iron and steel, cement and other building materials, including notably glass.[4] In these industries, the GCC countries have not yet achieved the status of global exporters, but are increasingly able to satisfy domestic demand with domestic production.

These industries have certain common characteristics that justify their successful development in the region. They are all capital-intensive industries, which is logical in a region notoriously short of labor and highly dependent on expatriates; and rich in financial resources. They are also in most cases technology and knowledge-intensive industries, which is in line with the aspirations of the countries in the region, and with their readiness to rely on imported know-how, for the time that is needed to develop a stronger national science base.

But a further key characteristic is that all of these industries are highly energy intensive. Oil refining, cracking and other petrochemical transformations, iron and steel, cement and ceramics, glass and plastic fabrication are all highly heat intensive transformations. Aluminum is somewhat different, as it requires large amounts of electric power for the smelting process, but electricity in turn is generated by heat. Hence the industrialization of the GCC countries is structurally characterized by high levels of energy consumption, and, for as long as the required energy is sourced from fossil fuels, CO_2 emissions.

Urban concentration and the role of services

Outside the industrial sector, further characteristics of Gulf development need to be included in the picture. A key one is the very high level of urban concentration in the major cities. Urbanization is a common feature of economic development in all countries, as the population systematically moves from lower-productivity agriculture to higher-productivity industry and service occupations in urban areas. But in the case of the GCC countries the process is somewhat more extreme, due to the starkly limited agricultural potential.

Indeed it may well be said that Gulf development is primarily a process of establishing a set of modern urban centers where some small villages existed in the past: centers mutually interconnected as well as tightly connected with

[4] These are discussed in: Hans-Georg Müller, "Other Major Manufacturing Industries" and Samir Pradhan, "GCC Building Materials Industry: Recent Trends and Prospects" in Giacomo Luciani (ed.), *Resource Blessed.*

the rest of the world, and competing with each other. It is not difficult to project this process into the future and visualize the formation of a crescent-shaped megalopolis running from al Jubail in Saudi Arabia to Dubai and Ras al Khaimah in the United Arab Emirates (UAE), and including Manama in Bahrain and Doha in Qatar. This string of urban centers will be connected by high speed rail in addition to air, and encompass a continuum of international airports served by some of the world's most dynamic airlines, major shipping and logistical facilities served by leading global logistics companies, offering sophisticated services in close association with all major global service centers. Urban concentration of a relatively large population in the harsh climatic conditions of the region is a further factor determining the relatively high level of energy consumption per capita.

A further crucial consideration relates to water supply. Already today almost all water available along the Gulf coast is derived from the desalination of seawater, but in the future it will be even more so. There is no way of procuring water in the region that would not be energy intensive.

Finally, specialization in services is also highly energy intensive. Transportation and logistics services are energy intensive, and everything connected to ICT is also dependent on reliable and abundant electricity supply.

The issue of energy supply therefore lies at the center of preoccupations concerning the sustainability of Gulf development. As the GCC countries will never accept that their development cannot be sustainable ("all that you see will revert to sand", to paraphrase the scriptural "Remember, O man, that you are dust, and unto dust you shall return", Genesis 3:19), the only relevant question in their eyes is how this kind of development—the only kind which is open to them—can be made sustainable.

Seeking sustainability with all available tools

The answer, not surprisingly, is that there is no single and simple solution. Sustainability must be sought through a complex strategy encompassing multiple policies and tools. The search for efficiency in energy use—stamping out waste, searching for the most performing solutions, maximizing synergies between different energy uses—is obviously a cornerstone, but the utmost priority must also be attributed to developing non-fossil primary sources, and the field should not be limited to one alternative.

It is obvious that the region has significant potential to exploit solar energy, while wind energy is a possibility along the Red Sea coast, less so along the coast of the Gulf. Other potential renewable sources should be carefully considered, even if currently not very promising, as surprises are not to be ruled out. However, it is not possible to rely exclusively on renewable sources, and nuclear energy has a fundamental role to play in the context of an overall effort to address the energy sustainability of Gulf development. Renewable sources and nuclear energy should be viewed as complementary, rather than substitutes.

What is the difference between nuclear, solar and wind energy? All three are primarily sources of electric power.[5] Nuclear energy can be viewed also as a source of heat, and this potential use will be discussed subsequently. But the main interest in the region is for electricity generation.

Power generation technologies and the variability of load

Electricity demand is highly variable, depending on the hour of the day, on the day of the week (whether working day or weekend), and even more so on the season of the year. As electricity cannot be stored, supply must be adjusted to demand instantly, lest the network collapses. Hence power sources are differentiated depending whether they are most suitable for providing base load (that is, that component of demand which is always or almost always active, independently of the hour, the day, the season), medium load (demand that is active frequently, but not always, such as demand that is repeated every day for a few hours, at all times of the year) and peak load (demand which is only seldom active, especially seasonal peaks).

There are economic and technical conditions, which determine whether a specific technology is more adapted to base, medium or peak load power generation. From the point of view of economics, base load is best addressed by technology that is capital intensive but minimizes variable cost (notably fuel), while peak load is best addressed by technology that minimizes fixed investment, even if variable costs are higher. Hence typically all large-scale thermal power plants—and certainly nuclear power plants—are best suited to provide base load.[6]

[5] Solar thermal can also be used for air-conditioning—at present a major component of electricity demand—through a process called absorption or adsorption. This technology has been implemented in several countries and is supported by, among others, the US Department of Energy. The International Energy Agency has a specific program devoted to "Solar Air Conditioning and Refrigeration" at http://www.iea-shc.org/task38/.

[6] That said, it has been argued that nuclear energy can be used for meeting peak load demand via the production and use of hydrogen. See C.W. Forsberg, "Nuclear

At the opposite extreme a gas turbine (single cycle)[7] power plant, which minimizes investment and has limited economies of scale, is best suited to provide peak load.

Technologically, care must be given to the time needed to start the power plant and reach maximum capacity—and the range is from nuclear, which is very rarely turned off, to a gas turbine, which is very easy to turn on and off and whose power output is easily modulated.

In this context, renewable sources offer the added difficulty that we cannot control when they are available or how much power they will be able to generate. We may predict—with increasing accuracy—whether clouds will cover the sky and prevent the sunrays from reaching a solar power plant; and we may predict (less accurately) whether the wind will blow and with what intensity, but we do not control either phenomenon. When power from renewable sources is available, it should be given priority over other sources, but other sources must be available for meeting demand when renewable sources are not.

What this means is that renewable sources such as solar and wind power "eat" into the base load and increase the requirement for elasticity of supply from the rest of the power generation options. This is not equally true of other renewable sources, such as hydro power or power generation based on biomass (as fuel), but the former is obviously not plausible for the Gulf countries, while the impact of the latter can only be limited.

The bottom line is that renewable energy sources are especially complementary with single cycle gas-fired power plants, or at most with combined cycle gas-fired power plants. However, this is only true if the gas-fired power plants are not integrated with water desalination, because the combination of gas-based power generation and water desalination creates problems for the flexibility of the former: as water desalination cannot be stopped, the gas-fired power plant must run even if there is no demand for its electricity output (for example on a day and time when the sun shines and lots of electricity might

Hydrogen for Peak Electricity Production and Spinning Reserve," ORNL/TM-2004/194 (Oak Ridge, TN: Oak Ridge National Laboratory, 2005); or, by the same author, "Sustainability by combining nuclear, fossil, and renewable energy sources," *Progress in Nuclear Energy*, Vol. 51 (2009), pp. 192–200.

[7] Combined cycle gas power plants are more capital intensive than single cycle plants, yet less so than thermal plants; and are more efficient, hence frequently considered as being adapted to providing base load, although they are also very flexible and may be used for medium and peak load.

be produced by solar plants). That this then makes the otherwise potentially efficient and cost-reducing option of combining power generation and water desalination impractical is a major problem for the Gulf countries.

The importance of load management

This discussion points to the supreme importance of load management for the future of energy supply in the GCC. All that is possible should be done to level out power demand, because large fluctuations impose huge added investment costs for power plants that are used only for a limited number of hours each year.

Load management entails a broad range of measures, some related to rational use of energy (for example, better insulation of buildings in order to reduce demand for air-conditioning), some related to creating incentives to shift consumption from the "full" to the "empty" hours (for example, running the dishwasher at night), and some, finally, related to encouraging the establishment of industrial activities that may be halted at times of peak residential and commercial demand (interruptible customers).

Water desalination may offer an opportunity here, because water can be stored while electricity cannot. There are two main technologies to desalinate sea water:

- The Multi Stage Flash (MSF) technology envisages use of heat to transform the water into steam and thus separate it from the salts contained in it. This technology is frequently coupled with large steam power plants (the heat can be recovered from the desalination process to be used in electricity generation). An alternative to MSF is the Multi Effect Distillation (MED) process, which similarly uses heat to transform water into steam, but is much less energy intensive than MSF and has been gaining ground in applications in the Gulf as well as elsewhere in the world.
- The Reverse Osmosis (RO) technology envisages pressuring seawater against a membrane that has the power of letting water pass through while retaining the salts. This technology absorbs electric power, needed to operate the pumps to pressurize the seawater.

Use of the RO technology can therefore open the door to a form of "power storage"—running the plants at times of low electricity demand and stopping them at times of peak demand. If the total MSF and RO capacity is sufficiently oversized to produce enough desalinated water even when the RO plants are not operated continuously (which of course will add to the investment bill) then some flattening of load is possible. Obviously, another possibility is to use power generated from renewable sources for RO water desalination—in which case the RO plants would only work when renewable power is available; however, this is certain to result in too many interruptions for the RO plants to work with an acceptable level of efficiency.

Some load flattening may also come from other industrial process—always at the cost of reduced capital utilization. Aluminum smelting is of particular importance because of its very high consumption of electricity. An aluminum smelter has several pot lines that are operated in parallel. Each pot line cannot be deprived of electricity supply for more than one or two hours, otherwise the metal in the pots will solidify and the pots need to be entirely discarded; however, it is possible to interrupt electricity supply in rotation to individual lines for shorter periods, effectively reducing the plant's total power demand for limited periods, which however might be very useful for meeting peak daily demand to coincide with the seasonal peak. The cost of doing so will be some reduction in the output of aluminum and in the efficiency of the plant.

Increasing the share of total electricity demand which is satisfied by power plants that are constantly (or almost) operating at capacity is therefore crucially important for the minimization of electricity costs. These plants are also most likely to be integrated with MSF/MED water desalination, and are essential to the survival of the GCC countries.

We summarize this discussion by underlining two key points: first the need to manage demand to maximize the share of base load over total electricity demand, and second, in case renewable sources are to become a significant component of total electricity supply, the need to compensate for their variable availability with enough standby capacity from other sources of a flexible nature.

Natural gas and power generation

Natural gas is the most important primary source for power generation in the Gulf, and it is also the source that is most easily combined with renewable sources. Thus *prima facie* the best solution for power generation in the Gulf might be to continue relying on gas. However, the GCC countries—somewhat paradoxically—face a shortage of gas availability.

Most of the gas produced in the GCC countries—except for Qatar—is in association with oil. In the past, such gas was deemed worthless by the companies exploiting the fields, and so flared. Saudi Arabia's industrialization drive started with the decision to invest in a gas-gathering network which would allow using the gas for industrial transformation: when this was started the opportunity cost for gas was zero (or negative, as in fact flaring gratuitously harms the environment).

A share of the gas produced in association with oil is reinjected in the fields in order to maintain pressure, and is therefore not available for other uses, at least for as long as oil is produced from the field; thereafter the field may come to be operated as a gas field. The volume of gas that is produced in association with oil and available for other uses is therefore conditioned by the requirements of oil production: the key determinants are how much oil is produced and what share of the gas needs to be reinjected. The bottom line is that gas demand for power generation and industrial transformation has outpaced supply, and all GCC countries except Qatar face a shortage of gas.[8] In some cases (Oman, Abu Dhabi) availability of gas for domestic use is also limited by the fact that some of the gas has been allocated on a long-term basis to liquid natural gas (LNG) export projects.

Resources of "non-associated" gas (gas that is not a by-product of oil production) are known to exist in several GCC countries (Qatar, Saudi Arabia, Abu Dhabi, Oman and more recently discoveries have been made in Kuwait), but with the exception of Qatar the cost of producing that gas is expected to be much higher than currently practised prices. Non-associated gas will be produced but it will need to be sold at much higher prices, making it much less attractive as primary source for base load power generation.

The paradox of regional gas scarcity is to a considerable extent the result of the GCC countries' failure to establish an integrated GCC gas grid and unified market. As mentioned already, non-associated gas is abundantly available in Qatar, but only one regional pipeline project has been implemented—the Dolphin pipeline that brings Qatari gas into Abu Dhabi and other UAE emirates and into Oman. The commercial terms for this export project are not very attractive for Qatar, and an expansion of the capacity of this pipeline has been ruled out, at least for the time being. Projects to export gas to Kuwait or even much closer Bahrain, or through Saudi Arabia, have never been able to take off.

[8] This paradox has been known for long and analyzed in multiple reports. Two recent ones: Wood Mackenzie, "The Arabian Irony—Increasing Imports to Maximise Exports," *Energy Markets Service Insight* (July 2010); Booz and Co., "Gas Shortage in the GCC: How to Bridge the Gap," (June 14, 2010).

Instead, Qatar has opted for the launch of massive LNG projects, which enable Qatari gas to be exported all over the world, including over very short distances, such as to Dubai[9]—which is an oddity. In the meantime, Qatar has imposed a moratorium on new export projects, and is more likely to expand the capacity of existing LNG plants rather than reverting to the consideration of regional pipeline projects. Hence, effectively—for the time being—the paradox of GCC gas scarcity is there to stay.

As gas becomes inevitably more expensive, its role in power generation will tend to be confined to where its intrinsic comparative advantage lies, which is in providing flexibility for meeting peak demand and compensating for the variable availability of renewable energy sources, to the extent that the latter will indeed become an important component of total power supply. Increasingly, base load will need to be satisfied with other primary sources.

Crude oil and oil products in power generation

Before continuing with the topic of nuclear power and its importance for the future of GCC development, a short paragraph needs to be devoted to the use of oil, which remains an important component of total power supply in Saudi Arabia and Kuwait especially.

In past decades—at least since between the early 1980s and the early 2000s—the opportunity cost of burning oil in power plants might have been considered to be small. In that period of time, quotas prescribed by the Organization of Petroleum Exporting Countries (OPEC) limited the oil production of Saudi Arabia and Kuwait. Saudi Arabia especially (not so Kuwait) had significant unused oil production capacity, and the direct cost of extracting oil when capacity is available and paid for is minimal. The oil that was burned in power plants could effectively not be exported, and cost very little to extract. The alternative would have been to leave it in the ground; but a barrel not produced today would, according to accounting standards, only be produced decades later, and its discounted value would therefore be close to zero.

[9] Shell and Golar LNG brought into service a Floating Storage and Regasification Unit in Dubai in December 2010. "Qatargas and Shell make first LNG delivery to Dubai," Shell press release (December 6, 2010). The attraction of the arrangement is that Dubai's demand for gas is in the summer, when electricity demand peaks, while most other markets for Qatar LNG have peak demand ahead and during the winter season in the Northern Hemisphere.

Today, however, the situation is completely different. Global oil demand is growing rapidly, and non-OPEC sources stagnate (the International Energy Agency recently concluded that non-OPEC production peaked in 2006). The GCC countries have the potential of increasing their oil production (including to a very limited extent Bahrain) and are all investing heavily in order to do so and to be able to increase exports. What this means is that today the opportunity cost of burning a barrel of oil for power generation is equal to the international price of the same. Almost everywhere in the world, crude oil (to be precise: heavy fuel oil, which is the least valuable product obtained from oil refining) has been abandoned as fuel for power generation, because of its high cost. In essence, the same calculus now logically applies to the GCC countries as well, and in fact the leaders of the GCC national oil companies have been voicing increasing concern for the prospect of increasing use of oil in power generation.

A further consideration relates to the characteristics of GCC refineries and the availability of residual fuel oil specifically. Power plants in Saudi Arabia use either straight crude oil (in which case also the lighter, more valuable molecules are burned for power generation) or residual fuel oil. The latter is an important component of the product slate from refineries that lack conversion capacity beyond the initial distillation phase. Some of the refineries established in the region in the 1970s and 1980s were indeed large but unsophisticated, and generated a large stream of residual fuel oil. The market for this product has progressively narrowed, as power generation shifted to alternative primary sources around the world, and its use came to be increasingly restricted to marine bunker (and more recently specifications for bunker have become tighter). Thus it may have been justified to burn residual fuel oil in power plants, as little alternative use for it existed.

However, significant investment has gone into upgrading some of the older refineries, and all of the new refineries are designed to maximize the production of valuable products and minimize residual. In most cases, refineries are now equipped with fluid catalytic converters (FCCs) which allow for maximization of the output of diesel and naphtha for transformation into petrochemicals, and generate very little residue or none at all.

Hence, both economically and technically, the use of crude oil or oil products in power generation is not a viable proposition for the future of GCC electricity supply. It is possible that diesel and/or LPG may continue to be used to generate power in remote locations isolated from the grid or to meet the pinnacle of peak demand; but base load electricity certainly cannot continue to come from oil.

The role of nuclear energy

There are thus only two alternatives effectively available to generate electricity for base load: coal and nuclear. Coal without carbon capture and sequestration (CCS) is the alternative that generates the most GHG emissions. With carbon capture and sequestration, the efficiency of coal-fired power plants is drastically reduced, and the cost of generated power is higher than in nuclear power plants—other things being equal. While technology may evolve in the future, and CCS may become more attractive, coal is not a very promising solution for the GCC today: nevertheless, Dubai and Oman are both considering the theoretical possibility of launching a coal-fired power project.

Thus nuclear is left as the only viable solution for base load power generation, and it is not a surprise that the GCC countries have manifested keen interest in acquiring a nuclear component to their power generation fleet.

As this chapter was being revised for final publication, a major accident occurred at the Fukushima nuclear power plant in Japan, following an earthquake and tsunami of extraordinary proportions. The Fukushima accident has led to a rethinking of the nuclear option in most countries, and the extent of its long-term impact on nuclear energy development may not be entirely clear for some time. Some countries, notably Germany and Switzerland, have taken steps to eventually terminate their use of nuclear energy, but mostly it appears that nuclear continues to be viewed as a valid option for reducing dependence on fossil sources. In the Gulf specifically, there has been no hint of a change in the interest for acquiring a nuclear component to the power generation fleet.

Nuclear energy in the GCC environment

In the rest of this chapter I will offer a critical analysis of economic, environmental, socio-political and technical conditions for nuclear power uptake, and argue that indeed it is especially suited to the region and very much complementary to growing reliance on renewable sources, assisted by gas.

Economic considerations

The economics of nuclear power is a contentious issue: estimates of the cost of the nuclear kilowatt hour (kWh) vary widely depending on the underlying assumptions. My purpose in the following discussion is to highlight the critical variables and discuss how these play in the GCC environment.

A nuclear power plant is a highly capital intensive operation, and the cost of investment dominates operating costs. The cost of fuel is smaller than for any other heat-based power generation alternative. The two crucial determinants of the cost of nuclear power are therefore the cost of the plant itself and the cost of financing.

The cost of the plant depends on the maturity of the technology and the degree of standardization reached in the design. Older technologies are more mature, and providers have acquired greater experience in the production of components and in the building process, allowing considerable reduction in costs. Opting for the latest design may improve the safety of the plant but adds to the cost. It is normally accepted that Abu Dhabi's decision to opt for a Korean supplier rather than the French Areva was determined by the fact that the Korean offer envisaged a mature and lower cost design, while the French company's offer envisaged its latest design, of which there are two specimens under construction, both very significantly over budget.

The trade-off between cost and technological advancement is an area where perceptions of safety may be expected to have an impact. Newer designs incorporate features that enhance safety, and the question is how much safety is considered necessary. The Fukushima plant was relatively old (its older unit was built in 1971) and was not designed to withstand a tsunami of the magnitude that struck the plant. There are no tsunamis in the Gulf, but one may expect that future orders, and the implementation of the UAE-Korean contract, will be influenced by relatively greater concern for safety, and relatively less for cost.

Plant capacity is also an important determinant of cost and potentially safety. Interest in smaller nuclear power plants has been increasing:[10] these plants can largely be manufactured in central manufacturing facilities and simply mounted locally, or even shipped as a complete finished product. It has even been proposed that plants may be sold or loaned inclusive of fuel sufficient for their entire working life, so that in fact they would act as long-term non-rechargeable batteries. Bundling the plant with its own lifetime fuel supply would also reduce the danger of proliferation, which is closely connected to the presence of plutonium in the spent fuel.

The revival of nuclear energy globally is likely to lead to increasing competition between providers (certainly the activism of the potential provider's country's heads of state and other high ranking government officials is a manifestation of intense competition) leading to declining costs.

[10] See the excellent information sheet by the Nuclear Energy Association on "Small Nuclear Power Reactors" at http://www.world-nuclear.org/info/inf33.html.

It should also be mentioned that much of the cost of nuclear power plants is due to civil engineering work, and the region has a sophisticated array of contractors that are well equipped to perform this part of the work and actively compete with each other.

Cost of financing is the other crucial consideration, and indeed all estimates of the cost of nuclear power normally specify the interest rate that is assumed to apply. The GCC countries enjoy large liquid balances and systematically engage in international investment, whether from "sovereign funds" or other sources (including central bank reserves, normally invested in very conservative low-yield bonds). For almost all the GCC countries, therefore, the opportunity interest rate or cost of financing of investment in nuclear power is what they can earn from international placements at the margin.

In addition, investment in nuclear power is a form of diversification and provides an asset that is under the direct control and in the jurisdiction of the investor, which increases its attractiveness.

Several GCC countries could provide financing for nuclear power without borrowing at all. Nevertheless, Abu Dhabi, which certainly has no need for borrowing, has opted to use some outside credit to finance its nuclear program with a view to improving control, thanks to the independent supervision of foreign lenders. This is a significant example of the pragmatic, non-nationalist attitude that frequently prevails in the region.

The physical environment for nuclear energy

The GCC states' physical and socio-political environment is also very conducive to the uptake of nuclear energy. Nuclear power plants have no extraordinary siting requirements, but certain characteristics are normally considered to be desirable, if not indispensible. The need for proximity to water, to provide for cooling, is common to all thermal power plants, whether fueled by coal, oil or nuclear energy. In the Gulf, base load power plants are ideally also coupled with water desalination, which absorbs some of the heat. Hence, almost all power plants (except for plants in Saudia Arabia serving the central region) are located along the coast.

Secondly, nuclear power plants are best located at some distance from major urban centers. This simple precaution serves the purpose of both limiting the potential impact of accidents and facilitating protection of the plant from intrusion. In most GCC countries, stretches of unoccupied coastline along which nuclear power plants can be located exist (if we consider the individual members of the UAE, Dubai may have a problem in this respect, as its coastline is now entirely urbanized; but at the federal level the problem no longer exists).

A third desirable characteristic for nuclear power plant is the absence of seismic risk, although countries that face significant seismic risk have successfully opted for a nuclear component to their power generation fleet (Japan and South Korea). The GCC countries have very stable geology and seismic risk is minimal.[11]

In comparative analysis with the rest of the world, the GCC countries may thus be considered to have a very favorable siting environment for nuclear energy, which is likely to translate into lower construction costs and good economics.

Socio-political considerations

A frequently heard objection to nuclear energy is that its implementation requires a long gestation period from the time the decision is made to the moment when a plant actually becomes available. This objection is not very compelling (it may be an argument for launching a nuclear program earlier rather than later) and in any case much of the long gestation period is due to the complexity of the authorization process and the possible opposition from local interests to the proximity of the chosen site.

But for the GCC countries the centralization of decision-making power is undoubtedly an advantage in this respect. Of course this is paid for with the lack of democratic scrutiny and debate, which is lamentable in principle, but the concentration of decision-making power facilitates the uptake of nuclear energy. It should be noted here that nuclear power generation, together with other large-scale and inherently centralized systems, is symbiotic with centralization of administrative power: it both requires and justifies the latter. Nuclear energy (indeed the management of the electricity grid) together with the oil and natural gas sector is a system that requires entrusting considerable power

[11] This is notably in contrast to Iran, where the seismic risk is high. For details see V. Pascucci, M.W. Free and Z.A. Lubkowski, "Seismic Hazard and Seismic Design Requirements for the Arabian Peninsula Region," paper presented at the 14th World Conference on Earthquake Engineering, Beijing (October 12–7, 2008); and Jamal A. Abdalla and Azm al-Homoud, "Earthquake Hazard Zonation of Eastern Arabia," paper presented at the 13th World Conference on Earthquake Engineering, Vancouver (August 1–6, 2004).

to an elite of individuals possessing technical and scientific knowledge and organized in corporate, hierarchical structures. Such "government of the competent" is very much in line with regional political discourse.

Know-how and technical readiness

This leads us to the next frequently heard objection, which is that GCC countries in fact do not possess the technical expertise needed to manage a nuclear energy industry—exactly in the same fashion as in the past it was argued that the region did not possess the technical knowledge to manage the oil and gas industry, or the petrochemical industry. The regional answer to the lack of sufficient numbers of technically educated nationals has been the pragmatic acceptance of dependence on technical expertise from abroad. This strategy, coupled with a constant yet measured effort at educating nationals and facilitating the transfer of technical know-how, has so far yielded very positive outcomes in several areas. The dependence on expatriate skills is not decreasing because the pace of demand continues to be greater than the increase in supply, but increasing numbers of nationals are being formed and their limited numbers have not become an obstacle to economic and industrial development.

The same approach may be expected to be applied to the development of nuclear energy. Initially, the task may be almost entirely entrusted to foreign providers, as the case of Abu Dhabi's deal with South Korea clearly indicates. In international comparison, this deal is quite unprecedented in the extent to which it pushes acceptance of reliance on foreign control of the industry—and it matches the decision to underwrite international obligations surrendering sovereign rights, such as the right to pursue uranium enrichment. It is not clear that other GCC countries will be willing to go as far as the UAE has gone, but the basic philosophy is likely to be similar.

Decision making processes for alternative energy sources

The energy requirements of GCC development therefore require the aggressive pursuit of all opportunities:

- Greater energy efficiency and reduction of waste;
- Management of demand to pursue reduced variation and increase the share of total consumption which is accounted for by base load;
- Greater reliance on renewable sources;
- Introduction of a nuclear energy component.

Progress in each of these directions depends on decision-making processes that are vastly different in nature. In some cases decisions need to be made centrally and by the government, but in other cases decisions are decentralized, in some cases truly atomized, and governmental control is far from complete. Hence, the final outcome cannot be fully determined *ex ante*.

Greater efficiency in energy consumption requires a decision making process that is extremely scattered, literally requiring the cooperation of millions of individual consumers and family units. Governments can influence the process through a variety of regulatory tools, most notably with respect to minimum product specifications for home appliances or vehicles, as well as through building codes and urban planning. In all cases the dictates of energy efficiency are likely to clash with other objectives, as in general better energy efficiency also entails higher initial cost (or additional investment to improve, for the insulation of existing buildings for example). Hence it is easy to predict that political authorities will not attribute exclusive importance to energy efficiency, and will attempt to strike some compromise.

Another tool to influence decentralized decision making is of course price signals—that is, increases in tariffs and fuel prices. While these are an indispensable part of the package, they are unlikely to be effective in isolation, because it is not realistic that the cost of energy use should be increased so much as to generate extensive uptake of energy saving solutions (these remain mostly non-competitive on a purely commercial basis even in OECD economies where energy is heavily taxed).

The management of demand may be pursued at the highly decentralized level of individual consumers or at the level of macro decisions based on specific technological solutions. The former requires progress in the direction of so-called smart grid solutions—installation of sophisticated meters capable of differentiating the tariff by the time of day or season, or even capability of remotely switching on and off specific consumption points or appliances/machines. This is both costly and difficult and has in essence not been implemented in any country yet. Macro decisions are more promising, and entail the planned reliance on interruptible process, such as RO desalination as discussed already. Grid integration at the regional level may also be of use because of the non-coincidence of daily demand peaks in different locations (in contrast, seasonal peaks are uniform).

Implementation of renewable energy solutions typically requires meso-level decisions, that is, a large yet not huge number of decisions to implement plants whose size varies from the very small to the medium-large. If we leave aside for a moment the smaller end of the spectrum (the installation of PV solar panels or solar air-conditioning on individual homes, for example), wind farms or concentrated solar power plants are typically medium size installations that require significant yet not huge investment. There may be relatively numerous investors who have the capability to undertake such projects, and numerous projects will need to be undertaken before their aggregate output makes a significant contribution to total electricity supply.

Hence the rapid uptake of solar and wind energy may be viewed as depending on the entrepreneurial and regulatory environment: will entrepreneurs be attracted to investing in these projects, will licenses and permissions be granted speedily so that a significant number of projects may be undertaken in parallel?

Obviously, the economics of the projects will play a fundamental role, and a prerequisite for successful deployment of renewable sources is that sufficient incentives are granted to investors (presumably in the form of feed-in tariffs). If the business is attractive, it is to be expected that investors will be keen to enter it, experience having shown that the region has a lively private sector that is constantly scouting for opportunities and quick to exploit one when it emerges. However, regulatory obstacles are likely to become a major problem, as obviously each individual project will need to be licensed, and the bureaucracy is very likely to be stifling.

In the end, it is a fact that electricity from renewable sources is structurally more decentralized than electricity from large-scale thermal plants, and the attitude of local authorities displays a strong bias for centralized control. It is easy to predict that renewable solutions will be popular for the supply of isolated installations in remote locations, which are either not connected to the national grid at all or insufficiently connected: in these cases, electricity from renewable sources may allow for a reduction of the use of diesel in small generators—the most expensive possible source. It is also conceivable that some licenses will be granted to a limited number of projects not too far from major urban centers, but it is much less likely that projects in remote locations requiring additional investment to be connected to the grid will be facilitated.

The advantage of nuclear energy is that it requires a small number of high-profile decisions, separated by long intervals of time. Once the decision is made, implementation is left to the bureaucracy which is then capable of performing the task rather well, with legitimacy bestowed from superior authority. The decision to build a set of nuclear power plants is one that will be taken by the top political authority in the country, but decisions about licensing medium sized plants based on renewable sources will need to be taken by lower level functionaries, and contradictions, unpredictability and procrastination are likely to be common.

Equally, macro decisions affecting load management are also likely to be made at the highest level, although with greater difficulty, as they may be more contentious. The key nexus is likely to be the link between power generation and water desalination, two overwhelming strategic interests of all GCC countries. Large thermal power plants will continue to be viewed with favor because they also allow for the lowest cost desalination.

Potential synergy between nuclear power and the optimization of hydrocarbons production

A further important consideration concerning the uptake of nuclear power in the region is the potential synergy between the nuclear energy and hydrocarbons production/transformation processes, with a view to achieving substantial decarbonization of the latter.[12]

The production of oil, its refining into petroleum products, and further transformation into petrochemical products require inputs of energy in various forms and at various stages. Upstream, the production of oil—especially "difficult" oil—may require steam injection or steam assisted gravity drainage (SADG). The treatment of Canadian oil sands requires substantial use of natural gas for heating and steam rising—in fact the process is known to entail very large CO_2 emissions. The same is true for Orinoco extra heavy oil deposits, while the production of oil from shale requires either heat or electricity. In the Gulf region most production has been of relatively "easy" oil, but increasingly challenging deposits are being brought into production, and the trend is bound to continue.

The refining process requires massive inputs of heat and hydrogen. Hence a part of the oil is burned in the refinery to generate the heat, and the hydrogen is normally obtained from some of the lighter products streams through reforming (separating the hydrogen from the carbon and combining the latter with oxygen, thus producing carbon monoxide or dioxide).

12 The discussion in this paragraph is based on ideas originally presented by Pierre-René Bauquis at the Mena Nuclear Energy Forum, Doha (November 10–11, 2008). A similar set of ideas is presented in Forsberg, "Sustainability by combining nuclear, fossil, and renewable energy sources."

Petrochemical plants similarly require large inputs of heat, hydrogen and electricity, all of which are normally obtained through burning fossil fuel. Refineries and petrochemical plants are some of the largest GHG industrial emitters, second only to power plants. The use of nuclear energy in lieu of fossil fuel to generate the heat or steam required by oil production, refining and transformation into petrochemicals may substantially reduce the carbon footprint of producing hydrocarbon fuels—which is not an indifferent consideration.

As for hydrogen needs, they may in perspective be satisfied through the electrolysis of water, which requires large input of electricity. The latter could also be provided for by nuclear energy (or renewable sources, if their intermittence can either be overcome or compensated for), further decarbonizing the process.

Of course this does not eliminate the GHG emissions generated by using hydrocarbons, notably in transportation. However, it must be recognized that the unsurpassed convenience of hydrocarbons in transportation use is linked to the extraordinary concentration of energy content per unit of mass that is the distinctive feature of hydrocarbons. No alternative fuel comes anywhere near providing so much energy per unit of mass, which means that any vehicle powered by any other fuel will have to devote much more of its transportation capacity to the fuel itself.

In fact, some experts believe that the best thing that we might do with CO_2 emissions is to add hydrogen to the molecule and produce synthetic hydrocarbons. This is possible and the technology is known: it is one that requires inputs of electricity and hydrogen, which again could be provided by nuclear energy with no carbon emissions.

The Gulf countries will increasingly evolve from being mere producers of crude oil and gas into being global purveyors of clean hydrocarbon fuels and petrochemical products. Along this path, their requirements of large inputs of heat, steam and hydrogen will hugely increase, considering the potential synergy between nuclear energy and the production of hydrocarbons and derivatives is of strategic interest to them.

Conclusion

I have attempted in these pages to propose a vision of the likely long-term development of the Gulf economies and explain why I believe that nuclear energy is an essential component of the picture. Only too frequently the Gulf economies are depicted as unsustainable and/or sources of major atmospheric pollution, but the issue of how these major problems might be overcome is not addressed. Renouncing development is not an option, and the best opportunity—if not the only one—that the Gulf countries have to lay the bases of development in the context of a globalized world is in their hydrocarbons endowment.

I have argued that the high energy intensity of the Gulf economies is to some extent the product of waste, but to a large extent a structural feature that is tied to their relative specialization. I have also argued that fossil sources are no longer sufficient to satisfy energy demand, and renewable sources face limitations to their potential role and speed of uptake. Nuclear energy is the only realistic solution to competitively satisfy the demand for energy and water in the region; and it is a promising tool to achieve decarbonization of the production of hydrocarbon fuels and petrochemical products, as well as the key to future recycling of carbon emissions through the production of synthetic hydrocarbons.

The potential military uses of nuclear technology are a source of concern and this must be addressed. We have an international framework for that, which is the Non Proliferation Treaty. This treaty clearly stipulates that all non-nuclear weapons signatory states have a right to access to civilian nuclear technology, including uranium enrichment. The credibility of the existing regime has progressively been eroded by its lack of universality and less than enthusiastic implementation on the part of the nuclear weapon states. Obviously the treaty must be revived and given a new lease of life, but this will not be achieved by erecting artificial obstacles to countries wishing to develop civilian nuclear programs.

At any rate, the nuclear weapons states that aim at tightening controls and limiting access to civilian nuclear technology have effectively lost the exclusive control of the technology which they enjoyed in the past. Thus attempts to prevent a development that is clearly in the national interest of the Gulf economies would only strain relations and, in the end, prove to be futile.

5

Saudi Arabia's Nuclear Future

Thomas W. Lippman

From the time the modern kingdom of Saudi Arabia was created in 1932 through to the end of the twentieth century, the engine of the country's development and modernization was money from the export of crude oil.

Lacking other resources, the kingdom relied on oil revenues to finance its astonishing progress in education, public health, housing, transportation and infrastructure. Starting from a base near zero, Saudi Arabia transformed itself in the span of one human lifetime into a fully modern country in every material sense, even as it clung to its cherished religious and social traditions. What Saudi Arabia wanted, it simply bought, including the imported expertise to build what the country needed and the manpower to do the work that Saudis lacked the skill or the desire to do.

Now in the twenty-first century the Kingdom's needs—and its economic aspirations—are different. Oil will remain the principal source of state revenues for the foreseeable future, but Saudi Arabia is no longer content—and soon will no longer be able—to function as a classic rentier state, living off the export of extracted natural resources and doling out fiscal sugarplums to its population.

The country has embarked on a hugely ambitious effort to transform itself into an industrialized economy, a powerhouse of petrochemicals, aluminum, communications and services that would rival South Korea or Norway in diversification of its economy, educational levels of its people and overall quality of life. The guiding document for this transformation is "Saudi Arabia: Long-Term Strategy 2025," promulgated by the ministry of Economy and Planning, which proclaims this "vision" for the coming years: "By the will of Allah, the Saudi economy in 2025 will be a more diversified, prosperous, private-sector-driven economy, providing rewarding job opportunities, quality education, excellent health care and necessary skills to ensure the well-being of all citizens while safeguarding Islamic values and the Kingdom's cultural heritage." It states flatly that "per capita GDP will double from its current level of SAR [Saudi Arabian riyals] 43,300 at the beginning of 2005 to SAR 98,500 in 2025 at constant 1999 prices."[1]

The objectives laid down in the master plan are to be implemented through plans for specific sectors such as labor, education and even agriculture. The plans may be unrealistic and certainly not all their goals will be reached, but the effort has begun. Oil money is financing it, mostly through massive infrastructure projects designed to entice private investors, but the key to implementation is electricity.

For the kingdom to pursue its goals while maintaining its citizens' quality of life, exponential increases in electricity-generating capacity will be essential over the next two or three decades. That reality is forcing the kingdom to make expensive decisions about what fuel will be used to produce the needed power. Rich in oil and gas resources as it may be, Saudi Arabia has concluded that in the long term the insatiable demand for electricity to power its growing industrial economy and support its escalating population cannot be met by burning hydrocarbon fuels; it will require the development of nuclear energy.

To outsiders the idea of nuclear energy in Saudi Arabia may seem illogical, as it does in Iran. Saudi Arabia has abundant energy resources: the world's greatest reserves of oil, the fourth largest reserves of natural gas, and unlimited potential for the development of solar power. Where is the need for nuclear?

The response from Saudi officials is straightforward and logically consistent. They say the country faces overlapping, interrelated demands on its resources that must be considered as a package. Saudi Arabia will need much more fresh water just to keep its population alive, and that water can come only from desalination of seawater, which requires huge inputs of electricity. Households

[1] This document is available in English at www.undp.org.sa/sa/documents/ourwork/pr/long_term_strategy_2025.pdf].

and individuals are using more power every year for appliances, gadgets and air-conditioners. The power-requirement projections for the planned new industries are daunting—and most of these industries will use natural gas for feedstock, reducing the supply of gas available to generate electricity. Burning oil in power plants cuts the amount available for export, an obviously undesirable choice in a country that derives nearly 90 percent of state revenues from oil exports. This all points to the desirability of nuclear power.

The quest for kilowatts

The Kingdom's population, now about 27 million, is growing rapidly. By 2050, it will have risen to almost 50 million, according to projections by the Population Reference Bureau.[2] If those Saudis to come resemble today's, they will habitually use more water than their environment can sustain; per capita water consumption in Saudi Arabia is the third highest of any country, behind only the United States and Canada. Because of the Kingdom's extremely limited natural resources of water—it is the world's largest country in area with no river or lake—providing water for that population will require a great expansion of the country's network of desalination plants, which are greedy consumers of electricity.

The scarcity of fresh water has forced Saudi Arabia to abandon its campaign to achieve agricultural self-sufficiency. That misguided campaign depleted the country's reserves of underground "fossil water," non-renewing reserves of water from some previous geological age that until recently were sucked to the surface to irrigate wheat and vegetable crops. Now, unless there is a complete redrawing of the Middle East's political map that would give Saudi Arabia access to the Nile or the Euphrates, the country has no choice but to multiply its energy-guzzling water desalination plants.

Moreover, the generation of electricity to meet current household, institutional and industrial demand, even without expansion, is already consuming so much oil and gas that the Kingdom's ability to export oil and provide gas to its existing industries is constrained. The country's planners have concluded that their ambitious aspirations for industrial development must be powered by some fuel source other than hydrocarbons.

Saudi Arabia already consumes the daily energy equivalent of 2.8 million barrels of oil, and in 2010 it became "the second-largest source of expected

[2] See various charts at www.prb.org/Datafinder/Geography.

global oil demand growth," behind only China, according to the International Energy Agency.[3] Put another way, the London-based Economist Intelligence Unit calculated that Saudi domestic energy consumption, which was the BTU equivalent of 114.6 million tons of oil in 2000, 153.2 million tons in 2005 and 198 million tons in tons in 2010, will rise to 341.6 million tons by 2020.[4]

According to the US-Saudi Arabian Business Council, a trade promotion group, "The power generation, transmission, and distribution industry in Saudi Arabia is currently one of the fastest growing and most lucrative in the Middle East. With one of the world's highest population growth rates and a rapidly expanding industrial base, the Kingdom has an ever growing need for electricity and power sources." With electricity demand growing by 7.7 percent a year, the Council calculated, Saudi Arabia "will need to significantly boost industry operations and expand its electricity grid by 300 percent over the next two decades."[5]

The public's faith in the ability of the Saudi Electricity Company to meet demand has been undermined by periodic power outages that have cut service to business and residences, especially in the Jeddah metropolitan area, and ignited occasional minor riots. People understand power shortages in Iraq; they do not understand them in peaceful, affluent Saudi Arabia, where reliable delivery of low-priced commodities and essentials is a fundamental part of the rulers' bargain with the politically disenfranchised population.

In November 2010, the giant state-owned oil company Saudi Aramco announced that it would build generating plants to meet its own needs rather than rely on the Saudi Electricity Company to supply the power. In reporting the development, the authoritative *Middle East Economic Digest* observed that "any extended period of production cuts because of power issues could result in the loss of millions of dollars of revenue. Aramco's decision to develop more power plants illustrates not just the importance of oil, but also a lack of faith in the ability of Saudi Electricity Company (SEC), the state utility, to provide it with the power it needs."[6]

[3] International Energy Agency (IEA) Monthly Oil Market Report (April 2010).

[4] Report at http://viewswire.eiu.com/report_dl.asp?mode=fi&fi=427089827.PDF &rf=0.

[5] US-Saudi Arabian Business Council, *A Business Guide to Saudi Arabia 2010* (Washington, DC: 2010), p. 103.

[6] *Middle East Economic Digest* (November 8–25, 2010) at http://www.meed.com/sectors/power/power-generation/saudi-aramco-to-go-solo-on-power/30791 23. article.

None of this comes as a surprise to the officials responsible for dealing with it. At an investment conference in Chicago in April 2010, Ali Saleh Al-Barrak, chief executive of the Saudi Electricity Company, said the country had 45 gigawatts of installed generating capacity as he spoke but would need 70 gigawatts by 2020 because of annual demand growth of about 7 percent. Planning to meet that demand is complicated by an immense disparity in usage between summer, when air-conditioners are running, and winter, when temperatures are comfortable.

In 2010 dollars, he said, at least US$80 billion will be needed to build the required generating facilities and transmission lines. "We have a long way to go to achieve what we want," he observed. He told an audience of prospective investors that generating plants with total capacity of 25 gigawatts are under contract or in varying stages of planning. All of them, he said, will be fueled by oil or natural gas or some combination of the two—that is, the need for more power is immediate and is being met with the resources at hand. Nuclear energy is in the future. When a member of the audience asked him what the fuel mix would be for the generating plants to come after those already on the drawing board, he replied, "We ask ourselves that every day."[7]

At about the same time as he was speaking, the International Energy Agency (IEA) noted that Saudi Arabia has sharply increased the burning of unrefined crude oil, rather than heavy fuel oil, in existing power plants. This reflects what the International Energy Agency (IEA) called Saudi Arabia's "limited success" in finding new sources of natural gas, and probably has some environmental benefit because crude pollutes less than heavy fuel oil. However, the agency said, it "effectively reduces that country's nominal spare [oil export] capacity. More importantly, perhaps, it leaves unaddressed the issue of runaway power demand, which has been largely due to highly-subsidized end-user prices."[8]

As that IEA report suggests, part of the reason for the skyrocketing consumption of electricity is that it is virtually given away. Saudi Arabia's citizens have little incentive to conserve because the price they pay is so low. Restaurants and public spaces are chilled to the point where people wear sweaters, and lights stay on all the time. Some increase in consumer rates is likely, but the Kingdom's rulers are reluctant to charge market rates for essentials such as

[7] Remarks made at the US-Saudi Business Opportunities Forum, Chicago (April 28, 2010).

[8] IEA Monthly Oil and Gas Report (April 2010).

electricity, water and gasoline because doing so would violate their longstanding bargain with the citizenry. Saudis have come to expect the subsidies as a way of benefitting in some small way from the country's oil wealth. A rate increase imposed in April 1990 was rescinded six months later because of widespread public opposition. Even if household rates go up, however, electricity use is unlikely to be reduced enough to make up for the increased demand from the country's factories and water desalination plants, which in turn is straining its natural gas supplies.

A natural gas shortage

The kingdom's acute need for more natural gas is a relatively recent problem. For many years, the gas that came up from the ground along with oil was simply flared, or burned away, because there was no demand for it. As the country electrified and began to industrialize in the 1970s, the folly of this waste became apparent, and Aramco undertook a massive project to capture the gas and distribute it to domestic consumers. The result was that gas-flaring emissions, which had been 38 billion cubic meters per year in the early 1980s, fell to 3.7 billion in 2010, according to figures compiled by the World Bank from satellite data. (As a comparison, this is slightly less than the figure for Libya, which produces only about 20 percent as much oil as Saudi Arabia).[9] But in the meantime, Saudi Arabia has committed itself to an industrial future built substantially on petrochemicals and other products that require natural gas as feedstock. As Oil Minister Ali Al-Naimi told the same investment conference in Chicago where SEC's Al-Barrak spoke, "We are entering a new phase of integration between the Kingdom's oil and industrial sectors. The early phase of developing basic petrochemicals is behind us, and we are promoting intermediate and specialty chemicals either by integrating the refineries with petrochemical facilities or investing to capture the full value added from our petrochemical potentials."

In the past two decades, the great bulk of Saudi Arabia's supply of natural gas has come from "associated" gas, a byproduct of oil production—the gas that in the old days was flared. Now the country's rapid population growth, urbanization, and commitment to petrochemical development have outstripped the supply of associated gas, creating a shortfall. As a result, oil that ideally would be available for export is burned in some power plants.

[9] Figures sourced at http:go.worldbank.org/G20AW2DKZ0.

An analysis of the Kingdom's natural gas demand by the US-Saudi Arabian Business Council found that Saudi Arabia has the highest per capita consumption in the world and its demand for gas will double by 2030. About one-third of all gas produced is used to generate electricity, 25 percent as fuel and feedstock for the petrochemical industry, and 20 percent for water desalination—all of which are expanding rapidly.[10]

For obvious reasons, the Kingdom is investing heavily in an effort to find freestanding gas reserves, with so far limited success. Oil Minister Al-Naimi told the Chicago gathering that "our investment in the exploration, production and processing of our gas resources has led to a 27 percent increase in reserves and a 60 percent increase in production from 5.3 to 8.5 billion cubic feet per day between 2000 and 2009. Today more than half our gas production is non-associated"—that is, it is produced by freestanding wells, not as a byproduct of oil wells—"and with the completion of the Karan, Manifa and Arabiah and Hasbah fields [the industry's] processing capacity will be more than 15.5 billion cubic feet per day by 2015."[11] He may be right, but independent analysts are skeptical.

An assessment of Saudi Arabia's gas needs by the US Energy Information Administration (EIA) in November 2009 was blunt:

> Rapid reserve development is necessary for Saudi Arabia's plans to fuel the growth of the petrochemical sector, as well as for power generation and for water desalination. According to Saudi Aramco forecasts, natural gas demand in the kingdom is expected to more than double to 14.5 billion cubic feet per day (Bcf/d) by 2030, up from an estimated 7.1 Bcf/d in 2007...However, natural gas production (estimated at 2.7 Tcf in 2007) remains limited, as soaring costs of production, exploration, processing and distribution of gas have squeezed supply, while an estimated 13 to 14 percent of total production is lost to venting, flaring, reinjection and natural processes, according to OPEC and other sources.[12]

Undersea reserves in the Gulf are expensive to develop and have an undesirably high sulfur content, the EIA report said, while explorations in the trackless "Empty Quarter" have shown little promise. Domestic consumption is higher than justified because the price was set in the days of associated gas, which was cheap; gas discovered a mile below ground after months or years of exploration requires correspondingly higher prices, EIA said.[13]

[10] US–Saudi Arabian Business Council, *A Business Guide to Saudi Arabia 2010*, p. 87.
[11] US–Saudi Business Opportunities Forum, Chicago (April 28, 2010).
[12] US Energy Information Administration, Country Analysis brief (November 2009).
[13] US Energy Information Administration, Country Analysis brief (November 2009).

As the EIA report noted, Saudi Arabia's ability to generate additional associated gas by producing more oil is limited by market demand for oil and by OPEC production quotas—that is, the Saudis cannot simply pump more oil to collect the gas that would come with it. A trade journal reported in October 2010 that a reduction in its OPEC quota had obliged Saudi Arabia to cut oil production and, as a result, gas production so much that there was an acute shortage of ethane, to a point where "stricter allocations of ethane have resulted in the Kingdom coming close to running out of feedstock resulting in short-term implications like disruptions to existing production."[14] Unlike some OPEC members that cheat on their quotas to export more oil than permitted by the cartel, Saudi Arabia—a founding member—generally adheres to OPEC strictures.

Here comes the sun

Over the long term, Saudi Arabia expects to meet a good portion of its electricity demand through the generation of solar power, and even to export solar power to its neighbors of the Gulf Cooperation Council (GCC) through an interconnected grid. In a comprehensive survey of the kingdom's long-term energy strategy delivered in 2009, Minister Al-Naimi said that Saudi Arabia "aims to be a leader in renewable energy production—specifically as the world's largest exporter of clean electric energy produced from our abundant sunlight." One of the scientific missions of the new King Abdullah University for Science and Technology, he said, will be to fund and guide a worldwide multi-institution effort to improve the technology of converting sunlight to electricity.[15] "Saudi Arabia aspires to export as much solar energy in the future as it exports oil now," Al-Naimi said a few months later at a kickoff ceremony for a multi-ministry research program aimed at fulfilling this goal.[16] But at the same time the demand for electricity is exploding for all the reasons already explained. Over the next several decades energy from renewable sources may eventually fill some of the gap, but in the meantime, when all these considera-

[14] Article at http://www.plastemart.com/Plastic-Technical-Article.asp?LiteratureID=1507.

[15] His remarks in text were distributed by the official Saudi Press Agency on March 16, 2009.

[16] P.K. Abdul Ghafour, "Solar energy initiative launched", *Arab News* (Jeddah: January 25, 2010).

tions are added up, the obvious conclusion would seem to be the development of nuclear power.

In theory at least, Saudi Arabia has accepted this conclusion and is committed to the development of nuclear energy. In 2006 the Kingdom joined all its partners in the six-nation GCC in declaring an intention to pursue the nuclear energy goal. In May 2008, President George W. Bush promised King Abdullah that the United States would "assist the Kingdom of Saudi Arabia to develop civilian nuclear power for use in medicine, industry and power generation." In announcing that commitment, the White House said that "the U.S. and Saudi Arabia will sign a Memorandum of Understanding in the area of peaceful civil nuclear energy cooperation. This agreement will pave the way for Saudi Arabia's access to safe, reliable fuel sources for energy reactors and demonstrate Saudi leadership as a positive non-proliferation model for the region."[17]

Two years later, in May 2010, the King issued a decree ordering the creation of a new government entity, the King Abdullah City for Atomic and Renewable Energy, to preside over the development of nuclear and solar power. "The City's goal," the decree said,

> is to contribute to sustainable development in the Kingdom by using science, research and industries related to renewable atomic energy for peaceful purposes...The City will be the concerned authority to fulfill national commitments pertaining to all agreements the Kingdom has signed or will sign regarding atomic and renewable energy. It will be responsible for supervising and monitoring all work related to the use of atomic energy and the resultant radioactive waste.

The work of the Kingdom's existing Nuclear Energy Research Center, a longstanding institution where the Kingdom's few nuclear scientists have worked on small-scale projects, would be transferred to the new organization, the decree said.

The King appointed himself as chairman of the new organization's Supreme Council and his half-brother Prince Sultan, the Defense Minister, as deputy chairman. The council's membership includes the ministers of Petroleum, Finance, Foreign Affairs, Health, Commerce, and Agriculture and Water.[18] Given the collective weight of that lineup, the King's decree appeared to rep-

[17] The text of the White House announcement is at http://www.america.gov/st/peacesec-english/2008/may/2008. The actual text of the joint memorandum itself has never been made public.

[18] Text distributed by the Saudi Press Agency (April 17, 2010).

resent a serious commitment to move forward with the development of a civilian nuclear energy program. But unlike neighboring Abu Dhabi, which faces energy requirements and constraints similar to those in Saudi Arabia and has awarded contracts to a South Korean firm to build nuclear power plants, the Kingdom is at least a decade away from construction. Only the most preliminary consulting work has begun.

Unlike many other countries, Saudi Arabia faces few obstacles if it decides to proceed. The government has the money, there is no political or environmental opposition, and there is plenty of uninhabited land where nuclear plants and nuclear waste repositories could be constructed. The biggest potential problem may be the identification of sites where nuclear plants could be constructed safely, a sensitive consideration in the wake of the Japanese disaster of 2011.

Some areas of the vast landscape that might otherwise be desirable may be geologically unsuitable, according to the Kingdom's leading expert on the subject, Dr Abdullah Al-Amri, professor of geophysics at King Saud University. He said his extensive surveys have found that the Gulf coast is "quite unstable seismically"—there is no site on the Gulf shore that would be geologically suitable because all the land there is sedimentary and subject to sinking and even liquefaction, as happened in some parts of Japan, in the event of an earthquake. The Red Sea coast is also unsuitable, he said, because there are indicators of potential volcanic activity there. Deep in the desert between Riyadh and Jeddah is a formation known as the Arabian Shield that is "one of the most geologically quiet places in the world" and thus could safely support nuclear plants, he said, but there is no water there to cool the reactors and spent fuel ponds. Large quantities of seawater would have to be pumped at least 30 miles inland, he said.[19] Saudi Arabia could afford to build the pumps and pipelines, but their vulnerability to rupture or sabotage would increase with each kilometer of distance from the source of the water.

In June 2010, a Finnish engineering and energy consulting firm, Poyry, announced that it had been selected by King Abdullah City for Atomic and Renewable Energy to "help prepare a draft of the national vision and high-level strategy in the area of nuclear and renewable energy applications" for the Kingdom. Company officials said at the time that their mandate was to help Saudi Arabia decide what its future mix of fuels would be and that nuclear

[19] Interview with Dr Abdullah Al-Amri conducted by author, Riyadh (April 12, 2011).

energy, if included, was a long-term proposition.[20] In a trade press interview after the announcement, David Cox, a Poyry executive based in Britain, was quoted as saying that the Saudi nuclear program could eventually include enrichment of its own uranium.[21] For reasons that will be explained below, that is highly unlikely.

A few weeks after that, three giants of the nuclear power industry announced they would team up to seek contracts in the new Saudi program. The three were Shaw Group Inc. of Louisiana, an engineering conglomerate; Toshiba Corp. of Japan; and Illinois-based Exelon Corp., a utility holding company that is the largest operator of nuclear power plants in the United States.[22] Shaw and Exelon are barred by US law from delivering nuclear materials to Saudi Arabia until the Kingdom reaches a formal nuclear cooperation agreement with the United States, but planning work can proceed and the announcement demonstrated that the industry believes Saudi Arabia is serious.

Nuclear power, nuclear weapons?

When President Bush committed the United States to helping Saudi Arabia develop commercial nuclear energy, the announcement generated surprisingly little controversy. The news went by almost as an afterthought because the American media's attention at the time was focused on the skyrocketing world price of oil and what Saudi Arabia was going to do about it, and because Bush had a reputation as perhaps the most pro-Israel of American presidents, who would do nothing to jeopardize that country's security. That casual attitude among the United States' public and its politicians is likely to change when the time actually comes for nuclear materials and equipment to be delivered to Saudi Arabia.

Saudi Arabia acceded to the Nuclear Non-proliferation Treaty in 1988, and in the small-scale work of its few nuclear scientists has complied with the requirements of the International Atomic Energy Agency (IAEA), the enforcement agency that monitors the nuclear activities of NPT states. Under the NPT's broad bargain, Saudi Arabia is entitled to acquire nuclear technology for peaceful purposes and to have the help of the recognized nuclear weapons states, including the United States, in doing so.

[20] Company news release (June 10, 2010).

[21] Articleathttp://www.utilities-me.com/article-60l-saudi-arabia-may-turn-to-uranium-enrichment.

[22] Glen Carey, Bloomberg News article (July 20, 2010).

In February 2009, Saudi Arabia and the IAEA reached a supplemental agreement in which

> Saudi Arabia undertakes, pursuant to paragraph 1 of Article III of the Treaty, to accept safeguards, in accordance with the terms of this Agreement, on all source or special fissionable material in all peaceful nuclear activities within its territory, under its jurisdiction or carried out under its control anywhere, for the exclusive purpose of verifying that such material is not diverted to nuclear weapons or other nuclear explosive devices.[23]

Nevertheless, in the minds of many Americans Saudi Arabia has been associated with terrorism and religious extremism since the events of September 11, 2001. Any use of the words "nuclear" and "Saudi Arabia" in the same sentence is bound to generate anxiety in Washington and Tel Aviv when the Saudis begin to move their ambitions from the theoretical to the real. Questions are sure to arise about nuclear weapons, about the potential diversion of nuclear materials, and about the security of nuclear material in the event of a regime change.

When Bush made his commitment to the Saudis, the Kingdom agreed to join the worldwide "Proliferation Security Initiative," a multi-faceted US program to combat the threat of nuclear proliferation, and the White House announcement listed multiple steps that Saudi Arabia had agreed to take to ensure that nuclear material would be kept secure, such as "enhance its accountability and physical protection of nuclear systems." But such commitments are of course based on the assumption that the government that made them remains in power and is dealing in good faith, an assumption Congress is unlikely to accept at face value.

In theory, the issues of nuclear power and nuclear weapons should be entirely separate, and often they are, as in Finland, for example, or in any other country that has one without the other. In the case of Saudi Arabia, however, any plan to develop nuclear power will inevitably raise questions about the parallel issue of acquiring nuclear weapons—largely because of the perception that Iran, a signatory state of the Non-Proliferation Treaty (NPT), is doing just that, but also because of what happened in North Korea, which was an NPT signatory state until one day it decided not to be.

Saudi Arabia is certain to be viewed differently from other countries to which the United States and some of its allies have been nuclear suppliers. Congress acquiesced in an agreement that allows the sale of US nuclear equip-

[23] IAEA Information Circular 746 (February 16, 2009).

ment to Saudi Arabia's neighbor Abu Dhabi, for example, because nobody views Abu Dhabi as a threat to anyone, but the Kingdom's strategic needs and ambitions are quite different from those of the tiny emirate. Saudi Arabia has responsibilities that Abu Dhabi does not, and thus is more likely to feel susceptible to a need to compete with Iran. The Kingdom is the largest military power on the Arab side of the Gulf and the only apparent counterweight to Iranian influence in the Persian Gulf barring Iraq; it is Iran's historic rival because of the Sunni-Shi'a divide, and it claims to have unique security responsibilities because of the annual Muslim pilgrimage to Mecca and its stewardship of the Holy Places of Islam. It cannot easily escape questions about its true intentions in pursuing nuclear energy, especially because of Israel's often-stated fear of a so-called "Islamic bomb."

The Chinese missiles

The possibility of a nuclear-armed Saudi Arabia has been on the table since the discovery in 1988 that the Kingdom had secretly acquired SS-20 missiles from China. These had always been seen previously as being potentially for use in a nuclear configuration, and because of their inaccuracy have no military value other than as a delivery system for nuclear or chemical weapons. The accidental discovery of this deployment brought an alarmed response from the United States, which basically forced the Saudis to accede to the NPT as the price of keeping the missiles. Previously the Saudis had said they would accede to the treaty only when Israel did so, knowing that Israel would not; accession was the price Riyadh paid to assuage US anger over the Chinese missile deal. If the matter had been put to a vote in the Kingdom, or if Saudi Arabia had had a legislature, accession to the NPT without any corresponding commitment from Israel might have been rejected, but—as always happens in Saudi Arabia—the decision was made without any public debate by the small circle of people who wield power in the Kingdom.

With its signature of the NPT, Saudi Arabia officially forswore nuclear weapons. To some analysts, however, acquisition of nuclear weapons could nevertheless make sense for Saudi Arabia because it is a vulnerable country, with great riches and a relatively small population, long porous borders and armed forces of dubious competence, surrounded by violence-prone and sometimes threatening neighbors. If the country was invaded, Saudi Arabia's armed forces could not retreat into a strategic heartland to regroup, as the Russians did when attacked by Napoleon and later Hitler, because most of the barren interior cannot sustain life. The Kingdom's greatest vulnerability lies along the Gulf coast, where the oil installations and desalination plants that are essential to the nation's survival could be easy targets for attack or sabotage. What would the Kingdom do if those facilities were under attack by a superior force?

These are troubling considerations for the Kingdom's leaders, but all these considerations have been part of Saudi Arabia's strategic equation for decades, and specialists generally agree there is nevertheless no credible evidence that Saudi Arabia has sought to develop or acquire nuclear weapons, or that it desires to do so—despite periodic alarming reports in the news media, especially in Britain. A 2001 study by the Carnegie Endowment for International Peace, for example, ranked potential proliferator states in the categories of "high risk," such as North Korea; "renunciations," such as South Africa and Ukraine, which had weapons programs but gave them up; and "abstaining countries" that could develop nuclear capability but chose not to do so, such as Taiwan. Saudi Arabia was not included in any category because it was not considered any risk at all.[24] Saudi Arabia refrained from the pursuit of nuclear weapons even knowing that Israel had them and that a hostile, threatening Iraq under Saddam Hussein was seeking to develop them. No evidence ever surfaced to indicate that Saudi Arabia was among the customers of the clandestine nuclear weapons market run by the Pakistani scientist A.Q. Khan.

What is different now is the possibility, perhaps even the likelihood, that Iran will soon have nuclear weapons. The people of Persia and the people of Arabia have been rivals since the dawn of Islam. Shi'a Iran today represents the most direct challenge to Saudi Arabia's primacy in the faith and to the legitimacy of the Al-Saud monarchy. Thus it raises the most important and most urgent question about the long-term security of Saudi Arabia and the Gulf: what will the Saudis do in the event that Iran develops, or is known to acquire, nuclear weapons and the means to deliver them? Will they try to acquire or develop their own nuclear arsenal, to balance Iran's power? Will they participate in a regional nuclear arms race? Will they refrain from the nuclear temptation and seek new security assurances from the longtime guarantor of their security, the United States? Or, most likely, do they not know what they would do because matters have not yet reached a point where they have to decide? Because of the opacity of Saudi Arabia's decision-making process,

[24] An updated "Proliferation Status Map" is on the organization's website at http://www.carnegieendowment.org/files/2009-global-prolif6.pdf.

there is no way to know for sure. What is knowable is that Saudi Arabia's effort to develop commercial nuclear power will not take place in a vacuum; as it progresses, Riyadh probably will not be able to avoid suspicious questions related to the Iran issue.

The fear among security analysts and proliferation specialists is not that Iran would actually use such weapons but that it would feel emboldened to throw its weight around, to the detriment of its Sunni neighbors. As Dr Lewis Dunn put it in a 2009 paper for the Naval Postgraduate School, "Among virtually all of Iran's neighbors, there is considerable concern that once in possession of nuclear weapons, the Iranian leadership could intensify its interference in those neighbors' domestic politics, step up support for extremist groups, and use the shadow of nuclear weapons to shape regional relations in a manner more favorable to it."[25]

As long ago as 2001, the strategic analyst Richard L. Russell argued the case for Saudi acquisition of nuclear capability. "It would be imprudent, to say the least," he wrote, "for Riyadh to make the cornerstone of [its] national-security posture out of an assumption that the United States would come to the kingdom's defense under any and all circumstances." It might be even more imprudent now, after the United States has overextended its security commitments in the Middle East and South Asia for a decade. "From Riyadh's perspective," wrote Russell, "the acquisition of nuclear weapons and secure delivery systems would appear logical and even necessary." Those "secure delivery systems," Russell argued, would not be aircraft, which are vulnerable to ground defenses, but "ballistic-missile delivery systems that would stand a near-invulnerable chance of penetrating enemy airspace"—namely, the CSS-20s, assuming that the obsolete behemoths have been upgraded by their Chinese suppliers and are still operational.[26]

According to the judgment of analysts at the RAND Corporation, which is shared by some others, "the advent of a nuclear-armed Iran would likely be perceived as an existential threat to Riyadh, possibly pushing Saudi Arabia to acquire its own countervailing deterrent."[27] Dennis Ross, the Middle East peace negotiations veteran who became President Obama's chief advisor on

[25] Lewis A. Dunn, "Strategic Reassurance If Iran 'Goes Nuclear,'" *Strategic Insights*, Vol. VII, No. 5 (December 2009).

[26] Richard L. Russell, "A Saudi Nuclear Option?", *Survival*, Vol. 43, No. 2 (Summer 2001): p. 70.

[27] Fredric Wehrey *et al.*, *Saudi-Iranian Relations Since the Fall of Saddam* (Santa Monica, CA: Rand Corp., 2009,) p. 98.

Iran, told a Washington audience in the spring of 2010 that "a nuclear-armed Iran would almost certainly precipitate a dangerous arms race in the Middle East, where states are already hyper-sensitive in their competition for regional influence and security."[28]

Powerful disincentives

When Secretary of State Clinton was preparing to visit Saudi Arabia in February 2010, the US Ambassador in Riyadh, James Smith, sent her a long message telling her what to expect in her conversation with King Abdullah. Smith said the King had told Gen. James L. Jones, White House national security advisor at the time, that "if Iran succeeded in developing nuclear weapons, everyone in the region would do the same, including Saudi Arabia."[29]

That may sound definitive, but it should not be taken at face value or as a statement of a considered policy that has been formally adopted by the Kingdom. It is by no means inevitable that the Saudis would seek their own nuclear arsenal. There are powerful disincentives to any attempt by Saudi Arabia to acquire or develop nuclear weapons, even apart from its shortage of technical and engineering capability. Any attempt by Saudi Arabia to develop or acquire nuclear weapons would undermine its security far more than it would enhance it.

First, it is unlikely that Saudi Arabia under its current rulers could develop a clandestine nuclear weapons program that would escape international notice; Americans are so deeply embedded in the Kingdom's security apparatus that they would surely pick up clues. And even aside from the fact that such a nuclear program would violate the NPT and place Saudi Arabia in the global nuclear outlaw category along with North Korea, the acquisition of warheads would encounter strenuous opposition from the United States and Israel. Having watched Washington's reaction to Pakistan's nuclear tests in 1998, the Saudis are well aware that US law requires economic and military sanctions against nuclear proliferators. And whereas Pakistan and India had friends in Congress, willing to help them escape the network of mandatory sanctions, Saudi Arabia does not. If an angry Congress cut off Saudi Arabia from future purchases of US military equipment and training, and Israel threatened a preemptive strike, its security would be diminished rather than enhanced.

[28] Address to Anti-defamation League National Leadership Conference, Washington, (May 3, 2010).

[29] http://cablegate.wikileaks.org/cable/2010/02/10RIYADH178.htm.

Moreover, confrontation and defiance are not Saudi Arabia's style; the Saudis' weapons of choice are cash and diplomacy. It is difficult to imagine the princes of the House of Saud deliberately positioning themselves as global outliers and inviting reprisals from countries capable of inflicting serious political and economic damage on them. With hundreds of billions of dollars of private Saudi capital and government funds invested in the United States, the Saudis would be ill-advised to risk an asset freeze.

Having committed itself to a long-term national development strategy that assumes full integration with the global economy, Saudi Arabia would undermine its own interests if it alienated the industrialized nations by turning into a nuclear rogue. Moreover, however strong its antipathy to Iran, Saudi Arabia does not face any current or foreseeable domestic or external threat to its security to which the use of nuclear weapons would be a relevant response. No country is capable of sustaining an occupation of Saudi Arabia, especially so long as the kingdom is under US protection, and nuclear weapons are useless against domestic troublemakers or border infiltrators. The benefits of acquiring nuclear weapons, if any, would be psychological and political rather than strategic, including a potential deterrent effect against reckless neighbors.

Any suspicion that Saudi Arabia was seeking nuclear weapons—or undertaking steps toward that end, such as building uranium enrichment facilities—would directly jeopardize all of the ambitious industrial development plans on which the country expects to base its economic future because, as explained above, those plans assume the availability of electricity that will probably have to be nuclear-generated, and Saudi Arabia cannot develop commercial nuclear energy unaided.

In pursuing nuclear electricity, the Saudis can be expected to seek a formal nuclear cooperation agreement with the United States similar to the one already negotiated by Washington and Abu Dhabi, the largest of the United Arab Emirates. Such an agreement, known as a 123 Agreement after the relevant section of the US Atomic Energy Act, would go well beyond the promissory memorandum given to the Saudis by President Bush in 2008; it would give the Saudis access to American nuclear materials and technology, and enable American companies such as Shaw and Exelon to supply them.

The Abu Dhabi agreement, which US officials view as a model for the region, requires the emirate to forswear nuclear weapons and prohibits it from enriching its own uranium or reprocessing the used fuel from its commercial plants to extract the plutonium content, either of which could be a prelude to the development of weapons.

As the Saudis are well aware, the likelihood that the US Congress would approve a 123 Agreement with Riyadh that did not contain similar restrictions is zero. Nevertheless, some officials have told potential American contractors that the Kingdom will not accept the Abu Dhabi model because of its restrictions on fuel cycle development. They said this is a matter of sovereignty—it does not reflect a desire to build enrichment capability but a reluctance to surrender the right to do so.

In theory Saudi Arabia could obtain commercial nuclear technology from countries other than the United States, as Iran has done with Russia. But the United States has many tools in its global anti-proliferation arsenal through such agencies as the Nuclear Suppliers Group, and in any case the Saudis, dependent on the United States for their national security, would want Washington's seal of approval for their nuclear energy program.

What if some unforeseeable event provoked a rupture between Riyadh and Washington? If the Saudis could no longer assume that the armed forces of the United States were their ultimate weapon against external threats, might they not wish to acquire a different ultimate weapon? The Saudi Arabian armed forces have never developed a coherent national security doctrine that could provide a serious basis for acquisition and deployment planning, let alone for a decision to acquire nuclear weapons. But to summarize the reasons why Saudi Arabia might be tempted to pursue such a course, it would find itself a rich but weak country with armed forces of questionable competence, outmanned by combat-hardened, truculent and potentially nuclear-armed neighbors, and no longer confident that it could count on its American protector.

Such a turn of events is theoretically possible—nervous Saudis often recall the example of Iran and the overthrow of the Shah—but highly unlikely. Saudi Arabia's princes are solidly entrenched and face no serious internal challenge to continued Al-Saud rule. Their alliance with the United States, shaken by events in the first decade of this century, is back on solid ground, despite neuralgic policy differences over events in the "Arab Spring" of 2011. Bilateral ties were bolstered anew by the Obama administration's decision to sell Saudi Arabia up to US$60 billion worth of advanced weapons and aircraft over the next ten years. That deal will require reinforcement of the American presence in the kingdom and of the US role in training Saudis to use and maintain the weapons. In addition, Americans are managing the training and deployment of a new 30,000-man Facilities Security Force to protect the oil industry and other critical installations, just as they have been doing with the Saudi Arabian National Guard since the 1970s. Now that the Saudi regime has sanitized itself, in Washington's estimation, of suspicion that it was complicit in extremist terrorism, there is no reason to anticipate a disruption of the alliance between Saudi Arabia and the United States. Any suspicion that Saudi Arabia was using its planned nuclear energy program to develop nuclear weapons would provide just such a reason.

Besides, the Saudis understand that Israel's reluctant acquiescence to the latest planned US weapons sale is based on Israel's calculation that it will not be the target of those weapons; Saudi Arabia is arming itself against many foes, real and prospective, but poses no direct threat to Israel. If on the other hand Saudi Arabia were suspected of having a nuclear weapons program, it would place itself near the top of Israel's target list, as happened when the SS-20s were discovered in 1988.

For all these reasons, the issue of nuclear weapons should be excluded from the discussion of Saudi Arabia's domestic energy needs. Those needs are urgent, but the stated commitment to include nuclear energy in the power-generation mix of the future does not necessarily mean it will come about because of what the Saudis themselves recognize as a wide gap between plans and implementation.

"The problem of implementation afflicts all levels of government," the master plan for 2025 notes. "Many projects and programs for which money has been approved are delayed. There are frequent cost and time overruns. Improving the capacity of the government to implement policies and programs has been called the mother of reforms. If this capacity is underdeveloped, then the implementation of all other reforms is likely to be adversely affected."[30]

Organizational capability is only one of the variables that could affect the pace and scope of Saudi Arabia's nuclear development. It is possible that Abu Dhabi's fast-moving program will result in a large surplus of generating capacity that will make it easier and cheaper for the Saudis to buy nuclear-generated electricity from the emirate than to produce their own. Other variables include oil prices, the results of natural gas exploration, technological innovation, the scope of conservation programs, progress in solar energy, geological evaluation, and the degree to which the anticipated industrial development actually occurs. The certainties are that Saudi Arabia in the next few decades will need much more electricity than it has now or has already contracted to develop, and that it has the money and the receptive domestic environment to pursue nuclear energy if such a decision is finally made. King Abdullah made

[30] See document referenced in note 1, section 3, part V.

his commitment with the creation of the new government unit charged with carrying out the development of nuclear energy, but it will be several years before any decisions are irreversible.

6

The Economic and Prestige Aspects of Abu Dhabi's Nuclear Program

Mari Luomi

The speed at which Abu Dhabi, the wealthiest emirate of the United Arab Emirates (UAE), established its civilian nuclear energy program in the late 2000s is something previously unseen in the region. After the six Gulf Cooperation Council (GCC) states announced in 2006 that they would consider nuclear energy, the monarchies' motives have been studied either in the frame of domestic energy security or in that of (non-)proliferation. On the basis of available information, the UAE's domestic energy security situation indeed seems to justify its choice. While the possibility of eventual proliferation ultimately depends on future political choices, and can therefore never be completely excluded, in the program's early years, Abu Dhabi has taken pains to convince the international community of its peaceful intentions. An important part of this endeavor was the formulation of the "UAE model," devised together with the Gulf state's key external ally, the United States, and employed by the local ruling elite in its pursuit of prestige in the eyes of regional and domestic audiences.

This chapter sheds light on the origins, motivations and characteristics of this "fast track nuclear program" and considers some of its main future challenges. It shows how a number of decisions and actions by Abu Dhabi's leadership in the early years of the nuclear program are not attributable to geopolitical ambitions or energy security alone. While the decision to pursue a domestic nuclear energy program was undoubtedly influenced by both economic and security considerations, the United States and France played a defining role in promoting and enabling the option for the UAE. This was most apparent in the swiftly signed bilateral cooperation agreements and the creation of the above mentioned nuclear model.

Also, in order to understand the UAE's nascent nuclear diplomacy, it should be acknowledged that prestige-seeking has become an important and distinguishable element of it. The chapter argues that, in the case of the UAE's nuclear program, prestige considerations are largely disconnected from the pursuit of military power. Prestige in this case is attached to non-proliferation-related values and symbols, in addition to notions of modernity, advanced development and cooperation, which bear important linkages to the general foreign policy orientation of the country.

After analyzing the economic motives and domestic emergence of the Abu Dhabi-led nuclear program, the chapter considers the role of external allies and the pursuit of prestige in molding the program in its formative years, from 2006 to 2010. Finally, the chapter shows how Abu Dhabi's apparent success in this massive energy project also has its shortcomings and weaknesses, which stem from the federation's political economy and external environment. While these weaknesses might not derail the nuclear program, they could bear important consequences for nuclear developments elsewhere in the region.

The nuclear choice

Experts have argued that three components must be present for a country to have a successful nuclear program, namely financial resources, scientific expertise and sustained political will.[1] In the case of Abu Dhabi, the owner of 7.3 percent of proven global oil reserves,[2] the first can by no means be seen to pose a significant constraint. The UAE, however, completely lacks the required domestic scientific and technical know-how, and as of 2008, it still lacked a regulatory structure. Until 2006, Abu Dhabi's lack of political will was also

[1] Mark Fitzpatrick (ed.), *Nuclear Programmes in the Middle East: In the Shadow of Iran* (London: International Institute for Strategic Studies, 2008), for example p. 55.

[2] BP, *Statistical Review of World Energy* (June 2010 edition).

seemingly self-evident, and explicable by its abundance of fossil fuel resources and the UAE's rather strong domestic defense capability and external security guarantees.[3]

As recently as 2006, the possibility of the UAE seeking a domestic nuclear energy production capacity was not considered by proliferation analysts as even a distant option, and a Saudi Arabian program, commonly held as more likely at the time, was also discarded by most. The military alliances and the regional presence of the United States were seen as the key reason for the GCC states not considering the option. Iran's potential acquisition of a nuclear weapon was seen as the only possible trigger for a change in resolve in any of the other Persian Gulf states.[4] Moreover, in the mid-2000s, domestic energy security concerns had not yet risen on government agendas in the GCC. What began unfolding in Abu Dhabi around 2006 was therefore against most expectations.

First, it should be clarified that in reality there is no such thing as the UAE's nuclear program. Despite cooperation agreements, regulatory structures and nuclear diplomacy being handled at the federal level, the nuclear program was initiated by Abu Dhabi and is set to remain so. Furthermore, given their financial and geographical restrictions, it is unlikely that the other emirates will seek to build nuclear plants in the foreseeable future. Although often understood as a federation, a more appropriate description of the UAE's polity is a confederation, such is the independence in local-level decision making of at least its most powerful and wealthy emirate, Abu Dhabi. Simultaneously, however, as the owner of over 90 percent of the UAE's fossil fuel resources, over which the constitution grants it sovereignty, Abu Dhabi's influence in federal decision making—not to mention federal foreign policy, which it effectively dominates—has arguably been growing in the aftermath of the 2008 financial crisis, which heavily impacted Dubai, and the growing energy crisis in the poorer northern emirates. Since the mid-2000s, Abu Dhabi's new generation of leaders has demonstrated increasing ambition to raise the emirate's regional standing through a number of initiatives, including a luxury

[3] A bilateral defense agreement from 1994 has enabled the US military to use the UAE's ports and hold troops at Abu Dhabi's Dhafrah airbase. Christopher Davidson, *Abu Dhabi: Oil and Beyond* (London: Hurst, 2009), p. 145; IISS, *Nuclear Programs in the Middle East*, p. 54.

[4] See, for example, Abbas Kadhim, "The Future of Nuclear Weapons in the Middle East", *Nonproliferation Review*, Vol. 13 (2006), p. 586; IISS, *Nuclear Programs in the Middle East*, p. 55.

tourism industry, a local airline, a French military base, an eco-utopia city, and a nuclear program.

Starting in 2007, the emirate of Abu Dhabi laid out the foundations of the world's fastest emerging civilian nuclear power program with the crucial help of external powers. Although it started without any prior domestic technical expertise, the government's political will and level of engagement were such that construction of the first of a total of four reactors is envisaged to begin in 2012, and they are expected to be operational as early as 2017. Even the Fukushima accident in 2011 did not cause any revisions to the plans.[5] Notably, since the official announcement by the UAE in 2008 that it would consider the nuclear option, transparency, non-proliferation and trust-building constituted Abu Dhabi's core internationally communicated intentions. As an indicator of the external credibility of its announcements, the UAE was already in 2008 declared by a leading international security institute as the most likely Arab country to first produce nuclear energy.[6] Around the same time, the UAE's main external patrons, the United States and the recent comer France, began drumming up support for Abu Dhabi's case by declaring its *modus operandi* as a "model for the world."[7] This chapter seeks to provide an explanation of what factors produced this outcome: how did the UAE's nuclear program emerge; why did the attitude of the Western powers turn from reluctant to eager; and what motives were involved on Abu Dhabi's side in its attempt to sell the program as a "gold standard"[8] to audiences in the region and beyond?

The economic motive

> [The] UAE's interest in nuclear energy stems exclusively from a desire to meet growing domestic energy demand in a commercially and environmentally responsible manner.[9]
>
> – Sheikh Abdullah bin Zayed Al Nahyan, Foreign Minister of the UAE, 2008

Iran's nuclear program is widely considered to have been the initial catalyst for the resurgence of interest in nuclear energy in the Middle East around the

[5] Correspondence with Christer Viktorsson, Deputy Director General for Operations, the Federal Authority for Nuclear Regulation of the UAE (April 2011).

[6] IISS, *Nuclear Programs in the Middle East*, p. 55.

[7] President Barack Obama quoted in Jay Solomon and Margaret Coker, "Oil-rich Arab state pushes nuclear bid with U.S. help," *Wall Street Journal* (New York: April 2, 2009).

[8] *Khaleej Times*, "UAE sets nuclear standard" (Dubai: September 26, 2010).

[9] Sheikh Abdullah bin Zayed Al Nahyan quoted in Emirates News Agency, "UAE

mid-2000s.[10] In December 2006, on Kuwait's proposal,[11] the six GCC states announced their intention to study the feasibility of a collective civilian nuclear program. Due to the timing and previous lack of interest, the move was perceived by most observers as a signal or an effort to create a strategic challenge to Iran.[12] This new interest in nuclear energy, however, has been seen and should be seen also as part of a broader rise of energy security on the global agenda since the early 2000s. Interest in nuclear energy resurfaced first in energy-importing countries as a consequence of rising oil prices and concern about climate change. Also playing a role was the relatively long period of time that had passed since disturbing accidents like Chernobyl and Three Mile Island. Moreover, assumptions regarding long-term oil prices began to change around 2005–6; the rising energy demand from emerging economies like China and India was starting to weigh in, and as a consequence, international projections were considerably revised upwards.[13]

Simultaneously, the oil-exporting Gulf monarchies found themselves suddenly confronted by enormous challenges in domestic provision of energy. This was a result of their economic and population booms in the 2000s, intensifying economic diversification efforts, highly subsidized electricity and water prices, and the consequently profligate consumption culture. The high domestic demand projections and lack of sufficient quantities of cheap natural gas to meet consumption were the main reasons that drove all the GCC monarchies except Qatar (with its natural gas reserves expected to last for up to three centuries)[14] to consider alternative energy sources. The high price projections

Government releases policy on Development of Peaceful Nuclear Energy", *UAE Interact* (April 21, 2008).

[10] See Sharon Squassoni, *Nuclear Energy: Rebirth of Resuscitation?* (Washington, DC: Carnegie Endowment for International Peace, 2009), p. 3. Other observers have noted that in many cases feasibility studies had begun long before: Meeting of the working group on *The Nuclear Question in the Middle East: Regional Perspectives*, Center for International and Regional Studies, Georgetown University, Doha (November 7, 2010). Chatham House rule applied.

[11] Economist Intelligence Unit (EIU), "Qatar: Energy Report", *EIU Industry Briefing* (December 1, 2009).

[12] See Joseph Cirincione and Uri Leventer, "The Middle East's nuclear surge", *New York Times* (August 13, 2007); Nicole Stracke, "Is the Gulf heading toward a nuclear arms race?" *Gulf in the Media* (Dubai: February 18, 2007); IISS, *Nuclear Programs in the Middle East*, p. 36.

[13] Meeting of the working group on *The Nuclear Question in the Middle East*.

[14] 280 years, at 2009 production rates. BP, *Statistical Review of World Energy*.

of oil made it a less attractive option for the exporters due to the higher opportunity cost.

The GCC joint study was conducted by the International Atomic Energy Agency (IAEA) and completed in November 2007. Multiple assurances of the civilian and peaceful nature of the program followed, and the UAE and later Kuwait pledged US$10 million to one of the international nuclear fuel bank proposals.[15] A report by the IISS (International Institute for Strategic Studies) in 2008 evaluated the GCC project, as outlined, as not posing a proliferation threat.[16] The collective plan, however, has not advanced since. Explanations for this arguably include the mutual distrust among the group's member states, the fact that sharing a nuclear program involves important security concerns that would be exacerbated by this distrust, and the inertia typical of multilateral processes compared to national projects.[17] Notably, not a single joint civilian nuclear energy program exists in the world to date. Clearly, one of the reasons why the GCC project has all but withered was the UAE's decision to take the traditional route, proceeding unilaterally.

Since the announcement of the program in 2008, the UAE government has predominantly used the economic frame to justify its relatively sudden interest in nuclear energy. Launching the UAE's nuclear policy in April 2008, Foreign Minister Sheikh Abdullah noted that nuclear energy represented "a commercially competitive and environmentally friendly option for the secure generation of electricity in the United Arab Emirates, particularly in light of projected future shortages of natural gas."[18] According to the local nuclear power company ENEC, the four factors supporting the choice were: economics, security of supply, environment and industrial development.[19] While the possibility of military motives influencing Abu Dhabi's choice can never be fully excluded, given the regional geopolitical setting, the economic argument deserves attention, for many reasons. Analysts who see military motivations behind the UAE's nuclear program have, in a rather populist manner, pointed

[15] Nuclear Threat Initiative, "IAEA nuclear fuel bank reaches funding target," *Global Security Newswire* (March 9, 2008).

[16] IISS, *Nuclear Programs in the Middle East*, p. 36.

[17] Ibid., pp. 35–6, 57. The IISS also reports on concerns among the five small member states over a "hegemonic overreaction on the part of Saudi Arabia" if it were left to take care of all oversight and licensing activities, as suggested by the IAEA (p. 36).

[18] Speech by Sheikh Abdullah bin Zayed quoted in Emirates News Agency, "UAE Government releases policy on Development of Peaceful Nuclear Energy."

[19] Emirates Nuclear Energy Corporation (ENEC), "Selection of Prime Contractor", Official ENEC Press Release (December 27, 2009).

to the vast fossil fuel reserves of Abu Dhabi as a proof of the program's predominantly strategic thrust.[20] In turn, those who have paid closer attention to the UAE's energy security situation understand that Abu Dhabi's available natural gas resources are relatively limited and its abundant oil resources are a valuable source of external revenue, which the government prefers not to waste in highly subsidized domestic power supply. Luciani has pointed out several economic reasons that have emerged since the mid-2000s to back the claim of both the UAE and the GCC in general: diminishing non-utilized crude oil production capacity, the growth of petrochemical industry (requiring gas as feedstock), the fact that nuclear energy is one of the cheapest sources of base load electricity (particularly if the price of carbon is factored in), as well as the considerations of opportunity cost mentioned above.[21] Furthermore, given that the UAE is, in Solingen's terms, an internationalizing model[22]—in which the leadership emphasizes economic competitiveness, openness and growth, and derives its legitimacy from this—the acquisition of a nuclear weapon would most likely entail untenably high political and economic costs for the ruling elite.

According to a high-level UAE official, the government first started considering nuclear energy in 2006.[23] However, in November 2006, the UAE Minister of Energy Mohammed bin Dhaen Al Hameli still denied any plans.[24] A high-level Western official has also observed that as late as January 2007 the government showed no real interest.[25]

[20] See for example: Joseph Cirincione, "Chain Reaction", *Foreign Policy*, web exclusive (May 2009) at http://www.foreignpolicy.com/articles/2009/05/06/chain_reaction.

[21] Giacomo Luciani, "The Gulf Countries and Nuclear Energy", *Gulf Monitor*, Vol. 6 (2007).

[22] Etel Solingen, *Nuclear Logics: Contrasting Paths in East Asia and the Middle East* (Princeton, NJ: Princeton University Press, 2007).

[23] Phone interview with Hamad Ali Al Kaabi, Permanent Representative of the UAE to the IAEA (November 2010), conducted by author.

[24] Solomon and Coker, "Oil-rich Arab state pushes nuclear bid with U.S. help"; Saifur Rahman, "Nuclear power for future UAE projects unlikely", *Gulf News* (Dubai: November 14, 2006). In 2005, the IAEA had approved a technical cooperation project aiming at assessing the "technical and economic feasibility" of a nuclear power and desalination plant in the UAE. Christopher M. Blanchard and Paul K. Kerr, *The United Arab Emirates Nuclear Program and Proposed U.S. Nuclear Cooperation*, Congressional Research Service (Washington, DC: July 17, 2009), p. 4.

[25] Seminar on "Energy and Climate Change in the GCC States", Foreign and Commonwealth Office, London (May 14, 2009). Chatham House rule applied.

Starting in 2006, an interagency energy planning committee in Abu Dhabi, consisting of the emirate's energy-related entities, was mandated to produce a more accurate estimate of the emirate's future demand of electricity and water.[26] Soon after, the mandate of this Energy Working Group was extended to the entire federation and included charting the feasibility of different technological options to meet future demand.[27] Simultaneously, since around 2007, France and the United States became increasingly involved in promoting nuclear energy for the UAE. The visit of President Sheikh Khalifa to France in July 2007 laid the foundation for bilateral nuclear cooperation. On President Sarkozy's initiative, the two presidents agreed to finally put into effect a 1980 agreement on the peaceful use of nuclear technology. An American nuclear consultancy also began working for a branch of the Abu Dhabi government since late 2007. In March 2008, after a rapid series of consulting, negotiations and bilateral agreements, the UAE's Supreme Council of Rulers approved a memorandum, submitted by Foreign Minister Sheikh Abdullah, which explained the motivations and principles of the country's potential plans to develop a peaceful nuclear energy program.[28]

In 2000–8, the UAE had enjoyed high economic growth rates of 6.8–9.5 percent per year.[29] As a consequence, owing to the small size of the national population (estimated at 900,000 in 2007),[30] a large influx of foreign workforce ensued, which led to a fast and unprecedented increase in the total population: from 3.1 million in 2000 to 5.6 million in 2006, and then to 5.1–6.0 million 2009, according to official estimates.[31] Moreover, stemming

[26] The committee included actors such as the power utilities and Masdar. Phone interview with Hamad Ali Al Kaabi conducted by author.

[27] Ibid.

[28] Emirates News Agency, "France to help UAE launch civilian nuclear program", *Khaleej Times* (Dubai: July 21, 2007); EIU, "UAE economy: Nuclear intent," *EIU ViewsWire* (May 28, 2008); website of the UAE Embassy in Washington, DC at http://www.uae-embassy.org/. (last accessed August 13, 2009).

[29] Low estimate (annual GDP growth on average): World Bank, *World Development Indicators* (April 2010 edition); high estimate (real annual GDP growth): EIU, *Data tool* (last accessed November 15, 2010).

[30] EIU, *United Arab Emirates: Country Profile 2008* (London: 2008), p. 12.

[31] Population statistics in the Gulf monarchies are often highly politicized, hence only estimates exist. Ministry of Economy of the UAE, *UAE in Numbers 2000* at http://www.economy.gov.ae/English/EconomicAndStatisticReports/StatisticReports/Pages/UAEinNumbers.aspx (last accessed November 28, 2010); Emirates Business 24/7, "UAE population to grow 6% in 2009," *UAE Interact* (May 19, 2009); Abdul-

from the rentier system and its allocative structures, which in the form of electricity and water subsidies extend to the entire resident population, consumption rates of these resources were among the highest in the world.[32] The growing petrochemicals industry and mature oilfields requiring enhanced oil recovery also consumed increasingly higher amounts of natural gas. As a result, domestic energy demand grew in the 2000s at unforeseeably high rates.

The government's policy paper, launched in April 2008 and known as the Nuclear White Paper,[33] includes a projection of the UAE's total demand growth by 2020. It states that the UAE's annual peak demand for electricity would be likely to rise by 9 percent per year, from around 13,000 megawatts (MW) in 2007 to over 40,000 MW by 2020. The paper also asserts that by this time, natural gas volumes available for electricity production would only provide 20–25,000 MW of total generation capacity, while renewables could supply only 6–7 percent of peak electricity demand,[34] in other words, 2,400–2,800 MW by 2020. The peak capacity gap remaining after gas and renewables would in this scenario thus be 12,200–17,600 MW, which equates to about nine to thirteen nuclear reactors with a capacity of 1,400 MW.

Moreover, the White Paper notes that despite their "logistical viability," burning crude oil or diesel for electricity production would "entail extremely high economic costs, as well as a significant degradation in the environmental performance of the UAE's electricity sector." Similarly, coal is discarded as a viable option, according to the paper, as environmentally unfriendly and involving "thorny issues related to security of supply."[35] A high-level UAE official later clarified that while coal at the time was commercially competitive, its carbon intensity alone was almost sufficient for withdrawing it as an option. An additional concern was dependence on imported coal, possibly amounting to "multiple shipments every week" through the geostrategically critical Strait of Hormuz,[36] which could be blocked by Iran, for example, in

---

lah Rasheed, "1.9 m increase in population since last official census," *Gulf News* (Dubai: October 6, 2009).

32 See WWF, *Living Planet Report 2010* (Gland: WWF, 2010).

33 Officially called the "Policy of the United Arab Emirates on the Evaluation and Potential Development of Peaceful Nuclear Energy."

34 Government of the United Arab Emirates (UAE), "Policy of the United Arab Emirates on the Evaluation and Potential Development of Peaceful Nuclear Energy" (April 20, 2008), p. 1.

35 Ibid.

36 Phone interview with Hamad Ali Al Kaabi conducted by author.

the case of an attack on its territory. In addition, the paper notes that, compared to the other options, "nuclear power-generation emerged as a proven, environmentally promising and commercially competitive option which could make a significant base-load contribution to the UAE's economy and future energy security."[37]

As a whole, Abu Dhabi's decision to choose nuclear energy, as it is presented to the world, hangs on a single paragraph of its fifteen-page Nuclear White Paper, as paraphrased and quoted above. Local officials have argued that the energy-related figures presented in the paper were based on an elaborate study, which, however, was never made public. Presumably, a large amount of decisive work and unpublished documents also include price estimates and detailed considerations on the dependencies and safety risks that nuclear energy creates. Here, too, a UAE official has affirmed that, while at the time of the evaluation it was impossible to calculate how much nuclear power in the UAE would eventually cost, international references and evaluations were used for an estimation that took into account factors that might vary if applied in the UAE's context. Scenarios for different energy mix options were also elaborated.[38] There are, however, at least two important questions that the White Paper does not address, which arguably should be addressed in public so as to increase the credibility of the economic argument: namely the methodology of calculating the maximum capacity provided by gas and renewables, and a comparative price estimate for exploiting domestic sour gas reserves versus importing liquefied natural gas (LNG) from Qatar or elsewhere versus nuclear energy. Also, the projections of the White Paper should arguably be revised in the form of an implementation update, to reflect the impacts of the global economic downturn on domestic energy demand.

Domestic post-2008 demand projections and international energy data, nevertheless, support many of the justifications presented in the White Paper: according to the Abu Dhabi Water and Electricity Company (ADWEC), the global economic crisis had little or no impact on electricity demand in the emirate. In 2010, it forecasted that local demand would rise from 6,300 MW (gross) to 19,000 MW in 2020 and 26,000 MW in 2030, representing roughly an annual increase of 10 percent in 2010–20 and 3 percent in 2020–30. Most

[37] Government of the UAE, Development of Peaceful Nuclear Energy, "Policy of the United Arab Emirates on the Evaluation and Potential Development of Peaceful Nuclear Energy", p. 1.

[38] Phone interview with Hamad Ali Al Kaabi conducted by author.

of this growth is expected to come from residential and commercial megaprojects and industrial sectors, indicating high economic growth and diversification expectations on behalf of the government, as set out in Abu Dhabi's urban plan for 2030.[39] Despite the 2008 economic crisis that hit Dubai particularly hard, official federal demand estimates for the 2010s were still high; in 2010, the UAE's Ministry of Energy was reported to estimate the federation's peak demand in 2020 at 33,400 MW,[40] which presents a significant decrease from the White Paper's projection (40,000 MW), but is still beyond what is said to be achievable with domestic gas supplies and solar alone.

Abu Dhabi owns relatively abundant natural gas reserves: around 3.4 percent of global proven reserves and 93 percent of the UAE's total reserves in 2009. In 2009, the UAE produced 1.87 trillion cubic feet (Tbc) of which it exported 248 billion cubic feet (Bcf), primarily to Japan, under long-term sales agreements. Practically all of Abu Dhabi's electricity is produced from gas, which is also used for water desalination and enhanced oil recovery. In 2007, however, rising domestic demand of energy drove the UAE to become a net gas importer, with 609 Bcf imported in 2009. This was partly due to the unplanned and unexpected growth and the late start in developing Abu Dhabi's domestic gas resources, a large share of which are ultra-sour and hence difficult to exploit.[41] The fact that most of Abu Dhabi's gas is associated gas is another major hindrance to its exploitation, partly due to production quotas prescribed by the Organization of Petroleum Exporting Countries (OPEC). According to government sources, approximately 50 to 60 percent of natural gas consumed domestically comes through the Dolphin pipeline from Qatar.[42] Despite existing additional capacity in the undersea pipeline, Qatar is not willing to export more to Abu Dhabi because of disagreements on the price

[39] ADWEC provides electricity and water also for the northern emirates and water for Sharjah. Keith Miller, Director of Planning & Studies Directorate, ADWEC, "ADWEC Winter 2009/2010 Demand Forecast", presentation at the MEED Arabian Power & Water Summit 2010, Abu Dhabi (March 30–1, 2010). Abu Dhabi's long-term urban plan is built on the assumption that the emirate's population will rise to 2 million in 2020 and 3–5 million in 2030. Abu Dhabi Urban Planning Council, *Plan Abu Dhabi 2030: Urban Structure Framework Plan* (2007), p. 45.

[40] MEED Arabian Power & Water Summit 2010,, "GCC power market report 2010: keeping pace with demand" (March 28, 2010).

[41] US Energy Information Administration (EIA), *Country Analysis Brief: The United Arab Emirates*, January 2011 edition.

[42] Phone interview with Hamad Ali Al Kaabi conducted by author.

(Abu Dhabi is seeking a "neighbor's price"), the moratorium Qatar has placed on its massive North Field due to concerns about the potential negative impacts of fast exploitation, and rising domestic demand of gas in Qatar.[43] In 2010, chronic power shortages, which had begun in 2007–08, still continued to plague the poorer Northern Emirates and Sharjah, which receive electricity from Abu Dhabi.[44]

Currently, 98 percent of Abu Dhabi's electricity is produced from natural gas.[45] In 2010, electricity production in the emirate was estimated to cost 18–25 fils (5–7 US$ cents) per kilowatt hour (KWh) while the industry, commercial sector and non-Emiratis paid 15 fils (4 US$ cents) per KWh and Emiratis only 3–5 fils (0.8–1.3 US$ cents) per KWh.[46] Although estimating the production cost of nuclear electricity in the UAE is still impossible, there are a number of international calculations that give a rough indication. According to an MIT study from 2009, the levelized cost of nuclear electricity in the US was on average 8.4 US$ cents per KWh.[47] A study by the Organization for Economic Cooperation and Development (OECD) projected costs for 2010 at 2.9–4.8 US$ cents per KWh in South Korea (which will build Abu Dhabi's first four reactors) and 2.9–11.5 US$ cents per KWh on average in the group's member states.[48] Hence, even though under the current subsidy regime electricity production would still not be able to recover production

[43] Justin Dargin, *The Dolphin Project: The Development of a Gulf Gas Initiative*, NG 22 (Oxford: Oxford Institute for Energy Studies, 2008), p. 48; EIU, *Qatar: Country Profile 2009* (London: 2009), p. 19.

[44] See Mahmoud Habboush, "Two-year timetable prepared to control power shortages in Northern Emirates", *The National* (Abu Dhabi: January 6, 2010).

[45] The rest is produced from oil. International Energy Agency (IEA)/Organization for Economic Cooperation and Development (OECD), *Energy Balances of Non-OECD Countries* (Paris: 2010 edition).

[46] Christopher Stanton, "Cheapest electricity since 1998 set for Abu Dhabi," *The National* (Abu Dhabi: October 19, 2010); Regulation and Supervision Bureau, *Electricity Tariffs for Large Users in the Emirate of Abu Dhabi* (November 2009), p. 3. The picture is of course not uniform: in 2010 a local newspaper announced that a massive 1,600 MW gas power plant, set to open in 2014, would produce electricity at 10 fils (0.5 US$ cents) per kWh. Stanton, "Cheapest electricity since 1998 set for Abu Dhabi."

[47] Massachusetts Institute of Technology, *The Future of the Nuclear Fuel Cycle: An Interdisciplinary MIT Study*, Summary Report (Cambridge, MA: 2010), p. 3.

[48] With different discount rates, including a carbon control cost of US$30 per ton of CO_2. World Nuclear Association, "The Economics of Nuclear Power" (March 9, 2011) at http://www.world-nuclear.org/info/inf02.html.

costs, if produced at a similar price or lower price than in the US or South Korea, nuclear energy would most probably be able to compete with alternative sources (domestic ultra-sour gas and LNG as the most important ones).[49] With an eye on the opportunity cost of using oil domestically, the IAEA feasibility study for the GCC from 2007 concluded that nuclear energy would be economically viable with oil prices of over US$50 per oil barrel (bbl). However, this would first require market-based pricing and the exhaustion of all associated gas reserves.[50] According to more recent IAEA estimates, nuclear power in the UAE will be economically feasible if oil prices stay above US$70 per bbl.[51] The UAE Foreign Minister Sheikh Abdullah, in turn, claimed in 2008 that nuclear energy would cost less than a third of electricity produced from crude oil.[52]

Abu Dhabi's known oil reserves are expected to last around 100 years with current production rates,[53] but as mentioned above, due to the current price projections and opportunity cost, oil has become an increasingly unattractive option for domestic energy supply. Notwithstanding the high uncertainties involved in future energy projections—for example, in 2010 the International Energy Agency projected prices in 2035 at between US$90 and US$140 per bbl—oil will probably still remain the global dominant fuel in the globally primary energy mix.[54] Also, it has higher carbon content than natural gas, similarly to coal. Greenhouse gas emissions are an issue Abu Dhabi is particularly aware of as a result of its alternative energy initiatives of the late 2000s, including Masdar and its hosting of the headquarters of the International

[49] Partly based on correspondence with Jim Krane (January 5, 2011).

[50] IISS, *Nuclear Programs in the Middle East*, pp. 35–6.

[51] Oliver Klaus, "Joining the nuclear club," *Wall Street Journal* (New York: March 15, 2010).

[52] Sheikh Abdullah bin Zayed quoted in Emirates News Agency, "UAE Government releases policy on Development of Peaceful Nuclear Energy." Oil price assumptions were not mentioned.

[53] Davidson, *Abu Dhabi*, p. 70. Peak oil theory supporters would argue that the UAE's choice of nuclear energy instead of oil is yet another sign of the oil producers running out of oil. See Ian Jackson, "Nuclear Energy and Proliferation Risks: Myths and Realities in the Persian Gulf," *International Affairs*, Vol. 85, (2009): p. 1158.

[54] As a central scenario. IEA, *World Energy Outlook*, Presentation to the Press (London: November 9, 2010) at http://www.worldenergyoutlook.org/docs/weo2010/weo2010_london_nov9.pdf; IEA, *World Energy Outlook 2010 Fact Sheets* at http://www.worldenergyoutlook.org/docs/weo2010/factsheets.pdf(last accessed December 15, 2010).

Renewable Energy Agency, as well as the perceived negative image impacts of its high per capita carbon footprint, often estimated among the highest in the world. Still, the casting aside of coal in one simple sentence in the White Paper leaves open questions, as the option remains on the table in other emirates of the UAE.[55] In November 2010, Dubai rather simplistically announced that its energy mix by 2030 would consist of 20 percent nuclear energy and 20 percent coal.[56] According to a local official, the UAE is currently looking into clean coal as an additional option to fill any remaining gap between total demand and supply after the nuclear plants become operational.[57] It should, however, be noted that there are large uncertainties associated with the commercial availability and viability of large-scale carbon capture and storage, which is needed for de-carbonizing coal power. Additionally, clean coal and dirty coal are no different in relation to transport and related concerns on import dependence and security of the supply chain. Finally, it should be noted that since the Emirate of Fujairah has its shoreline on the Gulf of Oman side, coal plants located there would avoid supply delivery through the Strait of Hormuz. The exclusion of coal by Abu Dhabi therefore clearly included energy dependence considerations not only relating to the maritime supply routes, but also in relation to its neighboring emirates.

As for renewables, in early 2009, the Abu Dhabi government took up a domestic goal of a 7 percent renewables capacity by 2020, which, with the above mentioned ADWEC figures, would translate into 1,330 MW.[58] In late 2009, this target was extended to the entire federation, the difference being that while the Abu Dhabi target is said to have been thought through and included in the emirate's investment plants, the federal level target only existed at the time of writing in the form of ministerial announcements.[59] While the 7 percent target has been considered by many renewable energy experts as too easily attainable relative to the available time and financial resources—and

[55] See "Tender for Dubai coal study gets strong interest," *Gulf News* (Dubai: November 14, 2010).

[56] Zaher Bitar, "Dubai plans peaceful nuclear power," *Gulf News* (Dubai: November 10, 2010).

[57] Phone interview with Hamad Ali Al Kaabi conducted by author.

[58] As of 2010, Abu Dhabi had built a 10 MW solar PV plant in Masdar City and was building a 100 MW solar CSP plant. Florian Neuhof, "Masdar City: Testing Times," *Utilities-ME.com* (November 8, 2010).

[59] Interview with UAE-based climate policy expert conducted by author, Abu Dhabi (October 2010).

thus not ambitious enough for an emirate aspiring to be the region's green energy leader—it has been often defended by local authorities as the first target to emerge from the Gulf region and as a show of goodwill and proactive behavior from a large oil producer. As a high-level official in the Ministry of Foreign Affairs noted in 2010, renewables in the UAE simply are not yet economically viable.[60] From an economic perspective there are a number of reasons why the deployment of renewables has been, and is expected to, remain relatively sluggish: the low cost of fossil fuels and high cost of solar energy; high electricity and water subsidies; and the lack of international obligations to curb domestic greenhouse gas emissions stemming from the UAE's developing country status in the United Nations Framework Convention on Climate Change. Feed-in-tariffs have nevertheless been planned by the Abu Dhabi government, which has since 2009 been promising to announce an energy policy that is expected to allow for establishing subsidy mechanisms for renewable energy.[61]

A second, albeit thin, energy narrative employed by the UAE government for justifying its choice has been the environmental explanation. According to the White Paper, nuclear energy was chosen because natural gas, the cleanest fossil fuel, is only a partial solution; renewables can only deliver a small share (and mainly serve for peak load supply); and because oil and coal are environmentally "detrimental." Worldwide, starting from the 2000s, the greenness of nuclear energy has commonly been employed as a supporting argument by its proponents. Environmental considerations have been mentioned by the UAE officials as factors that influenced the decision, and nuclear energy was mentioned by Foreign Minister Sheikh Abdullah at the UN's Cancún climate change conference of 2009 as part of the UAE's contribution to global climate change mitigation.[62] Although according to a UAE-based expert the diplomatic dimensions of climate change and nuclear energy are very distinct in the government's thinking, some officials have suggested that awareness of the

[60] Executive Seminar of the Zayed Diplomatic Academy entitled "Beyond Copenhagen: Environmental Diplomacy at Crossroads", Zayed University and the Ministry of Foreign Affairs of the UAE, Abu Dhabi (May 4, 2010). Chatham House rule applied.

[61] Chris Stanton, "Abu Dhabi to boost alternative energy", *The National* (Abu Dhabi: July 10, 2010).

[62] Sheikh Abdullah bin Zayed Al Nahyan, Minister of Foreign Affairs of the UAE, statement in the joint high-level segment of COP and CMP, 16th Conference of the Parties, UNFCCC, Cancún (December 9, 2010).

potential future international commitments the UAE might have to assume as part of a global climate change agreement (that is, cutting its domestic emissions) is a factor that is increasingly taken into account while evaluating particularly Abu Dhabi's future energy mix.[63]

The role of external supporters

> The agreement supports and advances long-term American interests in the Middle East region, while simultaneously allowing U.S. companies to compete for new business that will result in creating high paying, skilled jobs.[64]
>
> – US-UAE Business Council update on the 123 agreement, 2009

Arguably, in a counterfactual scenario, had the nuclear option not been made available by its external allies, the UAE's leadership would most likely never have made the choices of establishing a civilian nuclear energy program and renouncing the right to domestic enrichment, granted under Article IV of the Non-Proliferation Treaty (NPT). With the nuclear option excluded by the major global suppliers in the face of the above described domestic energy challenges, Abu Dhabi would arguably have resorted to a mix of other sources, ranging from imported market-priced natural gas (LNG) and/or coal, possibly crude oil, and a limited renewables capacity. It can be argued that the energy security outcome from the UAE's perspective would in this case have not been radically different (although its greenhouse gas emissions would have kept soaring). Given this, the role of suppliers of nuclear technology and the political support of major Western powers in the fast and so far successful emergence of Abu Dhabi's nuclear program should be examined more closely.

Arguably, since the mid-2000s, denying developing countries nuclear technology for fear of proliferation has no longer been the global norm—with of course apparent exceptions like Iran.[65] Despite concerns about proliferation and even safety, as Squassoni has noted, it has been the "advanced nuclear states [that] have largely generated the global surge of enthusiasm." The administrations of Presidents Bush and Obama have been eager promoters of

[63] Interview with UAE-based climate policy expert conducted by the author..

[64] US-UAE Business Council entitled "US-UAE. Agreement for Peaceful Nuclear Energy Cooperation: Advancing Mutual US and UAE Security and Economic Interests" (April 8, 2009).

[65] Jackson, "Nuclear Energy and Proliferation Risks", p. 1158.

nuclear energy. The US-initiated Global Nuclear Energy Partnership from 2006 and a joint declaration by the US and Russia to support nuclear energy in developing countries were major catalysts in this trend. The approach of the Nuclear Suppliers Group has been to promote "nuclear energy for all but only limited nuclear fuel cycles for most." The Group has been encouraging countries to voluntarily forgo domestic enrichment and reprocessing as "a confidence-building measure."[66] Here, the UAE is the prime example of a successful outcome from the suppliers' perspective.

Even before the announcement of the UAE's White Paper in 2008, major global suppliers of nuclear technology and know-how sought to secure their stakes in "the nuclear race" to the UAE. France, the US, Japan, the UK, and the newcomer South Korea, among others, signed memoranda of understanding (MoU) or cooperation agreements with the UAE government between 2008 and 2010. The main motive for the suppliers undoubtedly was the tens of billions of US dollars worth of trade opportunities expected during the coming decade and beyond. At the time it was speculated that the winner of the first four-reactor contract would also take home any other future deals. Furthermore, due to the downturn in the popularity of nuclear energy in the 1990s and 2000s, many countries with expertise that had not been able to build in their home countries, such as the United States until early 2010, were anxious to break into new markets.[67] For the countries at the receiving end, of course, this situation created benefits: at the first stage, as many as nine contractors were competing for the UAE deal, which consisted of designing, building and helping to operate four nuclear reactors. From the three finalists from which bids were requested—French Areva, US-Japanese GE-Hitachi and Korean KEPCO—Abu Dhabi is said to have chosen the South Korean consortium, a newcomer in international exports, principally because it was able to provide the cheapest offer. The expected reliability of the Korean schedule was another key criterion—one on which Areva failed to convince due to its problems with the Olkiluoto-3 reactor in Finland.[68] Signaling Abu Dhabi's increasing confidence, the two other final bidders' capability to provide the UAE with geopolitical support did not influence the decision; on the

[66] Squassoni, *Nuclear Energy*, pp. 1, 63.

[67] Seminar on "Energy and Climate Change in the GCC States."

[68] Steve Kidd, "The UAE reactor—why is it so significant?", *Nuclear Engineering International* (Kent: May 17, 2010); Joseph George, "Bids invited for N-power", *Emirates Business 24/7* (June 23, 2008).

contrary, the decision was presented in terms of technological engagement—the Koreans were lauded by the Emirates Nuclear Energy Corporation's CEO for having "dedicated a highly experienced team... [and] shown a serious commitment to transferring the knowledge gained from Korea's 30 years of successful nuclear industry operation."[69]

The Emirati-Korean deal has further benefits for both sides. In addition to the US$20.4 billion deal with KEPCO, South Korean companies are expecting another US$20 billion over the reactors' lifespan in maintenance, service and fuel supply contracts.[70] For KEPCO, the victory in December 2009 marked a long-awaited breakthrough into international markets. After the Korean victory, developments in the area of bilateral cooperation have confirmed the expectation that the contract would increase trade deals in other sectors as well: at the end of 2009, Mubadala and the Korean Ministry of Knowledge Economy announced an agreement to promote joint projects between Masdar and Korean renewable energy and technology companies.[71] In October 2010, Abu Dhabi awarded a Korean-Japanese consortium a gas power plant deal worth US$1.4 billion; in March 2011, the UAE awarded South Korea's President Lee Myung-bak a US$500,000 Zayed Prize for Global Leadership in Environment; and the two countries announced a MoU under which Abu Dhabi would commit to exporting over 1 billion barrels of oil to South Korea.[72] The nuclear project is estimated to require cooperation for close to 100 years,[73] indicating that this was only the prelude to enhanced bilateral relations between the two countries.

The suppliers' eagerness to provide the UAE with the needed technology was undoubtedly linked directly to the strong non-proliferation safeguards the UAE attached to its policy. Everything was done by the book, and beyond. The process through which the UAE's nuclear policy was developed included multi-stakeholder consultations at both domestic and international levels;

69 ENEC, "Selection of Prime Contractor."

70 David Adam Scott, "South Korea's Global Nuclear Ambitions," *The Asia-Pacific Journal*, Vol. 1 (2010).

71 Daniel Bardsley, "Masdar to work on technology with Koreans", *The National* (Abu Dhabi: December 28, 2009).

72 Stanton, "Cheapest electricity since 1998 set for Abu Dhabi"; Vesela Todorova, "Korean leader wins Zayed Prize", *The National* (Abu Dhabi: March 15, 2011); *Reuters*, "S. Korea secures Abu Dhabi oil deal" (March 14, 2011).

73 Khaldoon Al Mubarak, Chairman of ENEC quoted in ENEC, "Selection of Prime Contractor."

after determining the viability of nuclear energy for the UAE, the government developed a set of guiding principles, later embedded in the White Paper, and conducted a study of international best practices in the industry, as mentioned above. Before formally endorsing the nuclear policy, the government engaged in consultations with a number of supplier countries (France, the US, the UK, Russia, China, Japan, Germany and South Korea) and the IAEA. In developing a high-level strategy for pursuing the development of the nuclear program, the UAE followed the IAEA's guidance and the agency's Milestones document.[74] A resulting "Roadmap document", elaborated in cooperation with a US consultancy[75] and shared with the IAEA, included a feasible schedule for construction and recommendations on the needed order of actions (with that of site selection highlighted) and the establishment of the required institutions and passing of the needed legislation.[76] What followed since was a well thought-out establishment of a federal framework, within which Abu Dhabi could build its four nuclear reactors and possibly many more, aiming at a world record speed. Table 6.1 presents a timeline of selected developments in the UAE's nuclear energy program until the end of 2010.

In October 2009 the UAE outlawed domestic enrichment and reprocessing, the first country in the world to do so.[77] It is perhaps impossible to establish whether the United States pressured the UAE to passing this law or if the move was more a voluntary one. What is clear is that both sides have gained from this choice: the UAE has been able to pursue its nuclear program with the full support of the West, and the US has been able to use the UAE's case as an incentive for Iran and other aspiring states.

Originally, the United States obviously had large commercial interests at stake. In addition to the potential job-creating effect, the UAE's program was commended in the US for creating an "important model for the anticipated global renaissance in nuclear energy."[78] The US-UAE Business Council expressed concern on the US side about failing (temporarily) to sign the agree-

[74] Hamad Al Kaabi (Permanent Representative of the UAE to the IAEA), "Challenges Faced by Developing Countries in Nuclear Power Deployment: UAE Approach." Presentation in an IAEA meeting (October 27, 2009).

[75] EIU, "UAE economy."

[76] Phone interview with Hamad Ali Al Kaabi conducted by author.

[77] Emirates News Agency, "UAE President issues law on peaceful uses of nuclear energy", *UAE Interact* (October 5, 2009).

[78] Danny Sebright, US-UAE Business Council, quoted in *World Nuclear News*, "USA signs 123 Agreement with UAE" (January 16, 2009).

Table 6.1: Timeline of the UAE's nuclear program

1995	Signing the IAEA Non-Proliferation Treaty
2003	Signing the IAEA Nuclear Safeguards Agreement
2003	Signing the IAEA Physical Protection of Nuclear Material Convention
2006	Commissioning the GCC joint nuclear feasibility study
April 20, 2008	Launch of the Nuclear White Paper
April 21, 2008	Signing a Memorandum of Understanding with the United States
June 22, 2008	Request for initial bids for building Abu Dhabi's first batch of reactors
August 7, 2008	Contribution of US$10 million to an international fuel bank initiative
January 16, 2009	Signing a nuclear cooperation agreement with the US
April 8, 2009	Signing the Additional Protocol to the IAEA's Safeguards Agreement
June 21, 2009	Signing a nuclear cooperation agreement with South Korea
July 2009	Receiving bids from three short-listed consortia for four nuclear reactors
August 1, 2009	Joining a number of IAEA conventions
September 2009	Announcement of a three-month delay in awarding the nuclear deal
October 4, 2009	Approval of federal legislation establishing the regulatory authority FANR and banning domestic enrichment and reprocessing
December 16, 2009	Coming into force of the nuclear cooperation agreement with the US
December 23, 2009	Official launch of Abu Dhabi's nuclear energy company ENEC
December 27, 2009	Selection of KEPCO to build the first four reactors, total capacity 5.2–5.6 gigawatts (GW)
February 22, 2010	Establishment of the International Advisory Board, chaired by Hans Blix
February 24, 2010	Signing a nuclear safety cooperation agreement with the US
March 17, 2010	Announcement to establish the educational Gulf Nuclear Energy Infrastructure Institute in Abu Dhabi
April 2010	Announcement that Braka (Abu Dhabi) would be ENEC's preferred site choice
May 2010	Granting of an environmental permit by the Environment Agency Abu Dhabi to ENEC at Braka
July 9, 2010	Granting of a site preparation licence by FANR to ENEC at Braka.
December 28, 2010	Submission of licence application for Braka units 1 and 2 by ENEC.

ment in a briefing of April 2009: "Without an agreement with the U.S., the U.A.E. government will pursue its civilian nuclear energy program with the assistance of other nations."[79]

The so-called 123 agreement between the two states was signed in January 2009.[80] It was endorsed by President Obama in May 2009, and the US Congress approved it in October the same year. The process was partly delayed by concerns on non-proliferation and rule of law and human rights violations, including a torture case involving Sheikh Issa bin Zayed Al Nahyan, son of the late ruler Sheikh Zayed, revealed in April 2009.[81] Representing the basic foreign relations dilemma of Western states dealing with Middle Eastern regimes that often have authoritarian political systems and opaque judiciaries, the episode showed yet again how existing alliances and significant business opportunities make governments turn a blind eye to existing problems.

On the UAE's side, a large part of the technology it sought depended on US technology, the availability of which was contingent on the signing of the 123 Agreement. Even KEPCO's winning APR-1400 reactor has been developed from a design by the American Westinghouse that still retains patents for some of its technologies.[82] Some have even suggested that the US nuclear umbrella and related security guarantees could have alone functioned as a sufficient pressure lever, had the US not agreed with the policy and legal model the UAE was pursuing.[83] In the final US-UAE agreement the non-proliferation conditions were described as going "beyond those required by the Nuclear Non-proliferation Treaty, the requirements of the Nuclear Sup-

[79] US-UAE Business Council, "US-UAE Agreement for Peaceful Nuclear Energy Cooperation."

[80] The United States and the UAE signed a Memorandum of Understanding (MoU) in April 2008. The signing of the 123 agreement was delayed due to the presidential transition in the US. The agreement receives its name from the section 123 of the US Atomic Energy Act.

[81] Mark Landles, "Despite torture video, U.S. and Emirates sign key pact," *New York Times* (May 21, 2009). Sheikh Issa was put under house arrest in May 2009 and acquitted in January 2010. Also see Marten Youssef, "Sheikh Issa verdict will stand," *The National* (Abu Dhabi: January 25, 2010).

[82] World Nuclear Association, "Nuclear Power in the United Arab Emirates" (July 2007)at http://www.world-nuclear.org/info/UAE_nuclear_power_inf123.html (last accessed November 15, 2010; Scott, "South Korea's Global Nuclear Ambitions."

[83] Dominic Moran, "The UAE nuclear debate," *ISN Security Watch* (Zurich: March 6, 2009).

pliers Group, and even current U.S. law" in including a binding legal commitment not to enrich uranium or produce plutonium.[84] The UAE's Ambassador to the IAEA has defended the country's freedom of choice by arguing that "the final decision was a UAE decision [regarding] what commitments were taken," and noting that not all countries' supply chains have US elements.[85] Nevertheless, it can be argued that due to the lack of domestic technological expertise and time, the only route to fast deployment for the UAE was the total outsourcing of the nuclear program while seeking to ensure by all means that there would be no political obstacles placed by major suppliers, most importantly the United States, stemming from concerns about proliferation.

Prestige as a foreign policy driver

By the adoption and implementation of these policies and strategies, UAE Government hopes to establish a model for other countries that do not possess nuclear energy programs to be able to develop national nuclear energy programs with the full support and confidence of the international community.[86]

– Anwar Gargash, Minister of State for Foreign Affairs of the UAE, 2010

Simply put, the UAE's choice is evidence of a confident nation that makes crucial decisions based on what is best for itself. [...] Nuclear energy is the future, and the UAE has taken a prime seat in that endeavor. The program and the way it was pursued are a textbook case of how to tap into the most valuable source of power without creating undue fears about ultimate intentions.[87]

– Anonymous opinion article in *The National*, 2009

A nuclear program can be aimed at energy production, weapons, or both. Each of these options arguably brings a series of benefits: nuclear energy brings energy security, nuclear weapons provide a military deterrence, and both options potentially carry important prestige opportunities. In the Emirati government's calculation, prestige derived from adhering to the norm of

[84] US House of Representatives Committee on Foreign Affairs, "Berman Introduces Resolution Approving U.S.-UAE Civilian Nuclear Cooperation", *News from the Committee* (Washington, DC: July 14, 2009).

[85] Phone interview with Hamad Ali Al Kaabi conducted by author.

[86] Speech by Anwar Gargash in "UAE's nuclear programme to boost power grid", *Gulf Times* (Qatar: September 22, 2010).

[87] Article presumably produced by the Ministry of Foreign Affairs, "After oil, now welcome to the nuclear family", *The National* (Abu Dhabi: December 28, 2009).

non-proliferation has held stronger advantages than prestige attached to the possession of a nuclear weapon or acquiring nuclear weapons capability.

As a result of the signing and ratification of the 123 agreement, there have been important foreign policy and geopolitical outcomes on both sides. The US has been able to use the UAE's nuclear program and the government's agreement to forgo domestic enrichment for seeking to demonstrate to Iran what it could achieve by renouncing domestic enrichment and making its program more transparent. For the Abu Dhabi government, in turn, the deal has provided an important prestige asset vis-à-vis domestic, regional and international audiences.

Prestige, it is argued, was not necessarily a defining motive for the Abu Dhabi government in its *decision* to build a domestic nuclear energy program. Domestic energy security and, possibly to some extent, the symbolic deterrent value that a nuclear energy program can offer vis-à-vis Iran are much more convincing as explanations. Rather, the prestige dimension has been a *consequence* of the UAE's choice to pursue its nuclear energy program in a certain manner. As Abu Dhabi has pushed forward with its plans to build four reactors, prestige has turned into an element of the government's new nuclear diplomacy. In some cases, it is argued, prestige derived from the nuclear program has transformed into a foreign policy end in itself.

Prestige is defined by the *Encyclopaedia Britannica* as "standing or estimation in the eyes of people" or "weight or credit in general opinion." Wylie defines the concept in the context of international relations as referring to "the high level of respect accorded to states by other actors in the international system" involving "a high standard either generally or with regard to a particular issue area." Hence prestige is received from others and is socially constructed.[88] While prestige has been criticized for being too vague for analytical purposes, O'Neill has aptly argued that this is not a reason for ignoring it.[89] Prestige is a well-accepted motive for domestic-level actions (individuals, firms and groups), and International Relations theories, particularly Realism, have recognized its role. However, the influence of prestige considerations on foreign policy—prestige as an end—as Wylie has argued, has not been studied enough. She has also made the important observation that since "prestige is often more important to states that lack established power," these may "place

[88] Lana Wylie, "Seeking Prestige: A Foreign Policy Goal," 48th Annual International Studies Association Convention Chicago (February 28–March 4, 2007), p. 3.

[89] Barry O'Neill, "Nuclear Weapons and National Prestige," *Cowles Foundation Discussion*, Paper No. 1560 (New Haven: Yale University, 2006), p. 1.

a greater premium" on it.[90] Prestige can be of different types, such as military, but also moral resolve: the NPT regime "appears to have shifted the norm concerning what acts grant prestige and legitimacy from the 1960s notion of joining 'the nuclear club' to the 1990s concept of joining 'the club of nations adhering to the NPT'", as Sagan has pointed out.[91] Therefore, normative prestige can be gained also from rejecting nuclear weapons. Countries with moral prestige can even become "moral authorities."[92]

Prestige can also be sought from portraying one's country as progressive, advanced or modern. In the GCC countries, examples such as Qatar hosting global sports events and branches of top American universities, Dubai constructing the world's tallest building or Abu Dhabi pledging to create a 50,000-inhabitant urban eco-utopia city in its desert outskirts (while, somewhat ironically, hosting a F1 race nearby) abound. With these events and achievements the Gulf monarchies hope to become and be seen as developed and modernized, the latter of which Dore defines as meaning "to transform one's society, or segments of it, in imitation of models, and under the influence of ideas, drawn from other countries which are seen as more advanced in some implicit scale of progress."[93]

Prestige has a domestic dimension as well: a national dress worn by an Emirati or a white golden Bentley certainly grants its owner a higher social standing in the eyes of others. The government can achieve prestige or a good reputation in the eyes of its citizens through evoking national pride, by for example developing the "Lamborghini of nuclear energy programs."[94] In the area of foreign policy, different international forums and high-level meetings are common venues for emphasizing the importance of a nation. Technology is another aspect: the pursuit of becoming the Gulf's "future energy leader"

[90] Wylie, "Seeking Prestige", pp. 4–5.

[91] Scott D. Sagan, "Why do States Build Nuclear Weapons: Three Models in Search of a Bomb", *International Security*, Vol. 21 (1996): p. 67.

[92] Lana Wylie, "Valuing Reputation and Prestige: Canadian Foreign Policy and the International Criminal Court," *American Review of Canadian Studies*, Vol. 39 (2009): p. 114; Barry O'Neill, *Honor, Symbols, and War* (Ann Arbor: University of Michigan Press, 1999), p. 196.

[93] R.P. Dore, "The Prestige Factor in International Affairs", *International Affairs*, Vol. 51 (1975): p. 195.

[94] On nuclear technology and national pride, see Jackson, "Nuclear Energy and Proliferation Risks", p. 1159. "Lamborghini" coined by Giacomo Luciani, Doha (January 2011).

or the first Middle Eastern state to operate a civilian nuclear power program serves as an important domestic legitimacy resource for Abu Dhabi's ruling elite.[95] Technological advancement of course also has important linkages to regional and international level prestige.

Prestige-seeking through moral authority, progressiveness or technological development is often not as expensive as acquiring an important military capability. In most cases however, it does not come for free. In the case of the UAE's nuclear energy program, setting the gold standard involved important time and financial resources of the country's foreign policy elite. Below, an examination of the ways and contexts in which the UAE's nuclear model was built will be used to show the importance of prestige considerations particularly at the foreign policy level.

The UAE seems to be greatly fixated on its reputation and image in the global stage, and particularly in the West. This tendency has become increasingly visible since the mid-2000s, with the economic boom and ensuing crisis in Dubai as well as the awakening and consequent fast growth in Abu Dhabi. Since around 2005, as a result of leadership changes, Abu Dhabi aggressively started pursuing both economic diversification and a more prominent role in the regional system. This, it is argued, made the concern for external image a central element of policymaking of the UAE, in which Abu Dhabi has tightened the lead particularly after the 2008 crash in Dubai.

For example, in the area of environmental sustainability, since around 2007, the federation's high per capita greenhouse gas emissions became perceived by the ruling elites in Dubai and Abu Dhabi as harmful for the country's image. As a direct consequence, the government set up a team to recalculate the country's ecological footprint and Dubai announced that all its buildings would have to comply with green building codes as of 2008.[96] Also, the Emirate of Abu Dhabi has been investing billions of US dollars in creating a "future energy" sector in its economy, while creating a more sustainable place to live in bypassing new building codes, pledging renewable energy targets, and choosing nuclear over coal. For Abu Dhabi, the victory in its campaign to host the headquarters of the newly established International Renewable Energy Agency (IRENA) in 2009 was an important recognition of the positive international image the emirate had gained through its alternative energy initia-

[95] See Christopher Davidson, *The United Arab Emirates: A Study in Survival* (Boulder and London: Lynne Rienner, 2005), pp. 70–87.

[96] Emmanuelle Landais, "Dubai to turn green in 2008," *Gulf News* (Dubai: October 24, 2007).

tives, linked to the Masdar Initiative, established in 2006.[97] Hosting the IRENA headquarters has also brought Abu Dhabi prestige and enabled it to extend its participation into other global energy forums, such as the Clean Energy Ministerials of the Major Economies Forum.[98] As the host of IRENA, Abu Dhabi has also sought to portray itself as a bridge between the North and the South.[99]

As for the foreign policy dimension of the UAE's nuclear program, there are various states and audiences from which Abu Dhabi could be seen as seeking recognition, ranging from the international to the regional level. By being a good global citizen that complies with all non-proliferation treaties and norms and engages with the "right" actors and relevant UN agencies, the UAE can seek to be perceived in the West and globally as a civilized, peaceful and modern country and a regional actor to be taken into account and trusted. Further, by doing more than enough, in other words outlawing enrichment and reprocessing, the UAE can seek to achieve moral authority by becoming a model student in the new nuclear nations' community.

The six key principles set in the UAE's Nuclear White Paper are a case in point:

1. Complete operational transparency;
2. The highest standards of non-proliferation;
3. The highest standards of safety and security;
4. Close coordination with the IAEA;
5. Cooperation with the governments and firms of responsible nuclear supplier nations;
6. A commitment to long-term sustainability.[100]

Whenever the nuclear program is brought up in high-level speeches by UAE officials for international audiences, the principles are mentioned, for

[97] See Mari Luomi, "Abu Dhabi's Alternative-Energy Initiatives: Seizing Climate-Change Opportunities", *Middle East Policy*, Vol. 16 (2009).

[98] "UAE to host second Clean Energy Ministerial in 2011", *Khaleej Times* (Dubai: July 22, 2010).

[99] Statement by Sheikh Abdullah bin Zayed Al Nahyan, Foreign Minister of the UAE, the General Debate of the 65th Session of the United Nations General Assembly, New York (September 28, 2010).

[100] Hamad Al Kaabi, *Challenges Faced by Developing Countries in Nuclear Power Deployment*; Emirates News Agency, "UAE Government releases policy on Development of Peaceful Nuclear Energy."

example in the UN General Assembly speech by Foreign Minister Sheikh Abdullah in 2010:

> The measures undertaken by the United Arab Emirates for building its program for the peaceful uses of nuclear energy in order to meet its growing energy requirements in the future, rely entirely on the highest standards of transparency, nuclear safety and non-proliferation, and [are] carried out under the supervision of the IAEA in full cooperation with the Agency and other friendly and responsible states with relevant experience in this field.[101]

The speech then continued:

> These principles have led us to adopt a series of commitments and strategies which are outlined in the policy document and which we believe constitute a useful model via which, not only the United Arab Emirates, but other non-nuclear states could pursue the benefits of peaceful nuclear power with the full backing and confidence of their domestic populations and the wider international community.[102]

Naturally, the security and proliferation concerns surrounding a Middle Eastern state pursuing a nuclear program have had a key impact on the way the program has been presented to international audiences. Transparency, safety and the peaceful nature of the program have all been part of the government's efforts to build trust and ensure the rapid implementation of nuclear energy in Abu Dhabi. Also, a point often highlighted by UAE officials is that excluding enrichment was not only a political choice, but also a cheaper option due to the high cost of enrichment and the so far relatively small scale of implementation plans in the federation.

However, what goes beyond all this is the model that has been invented along the way. While it is clear that the United States had a key role in envisaging the UAE's model and how it should be sold to Middle Eastern states aspiring to nuclear energy and/or weapons, on the UAE's side, a more multi-dimensional discourse is distinguishable. Messages are aimed at multiple audiences, ranging from Emiratis, through GCC countries, other Arab states and Iran, to the US/West and international audiences in general. These messages are grouped below into seven areas of "exceptionalism" according to the types of prestige the UAE is argued to seek from its "model nuclear program":

101 Statement by Sheikh Abdullah bin Zayed Al Nahyan, the General Debate of the 65th Session of the United Nations General Assembly.

102 Sheikh Abdullah bin Zayed quoted in Emirates News Agency, "UAE Government releases policy on Development of Peaceful Nuclear Energy."

1) *Model for developing countries*: As explained above, Abu Dhabi wishes that its program would establish a gold standard for all new countries seeking civilian nuclear energy: "the UAE is leading the new nuclear renaissance," as expressed by an UAE official.[103]
2) *Showing the weakness of the Iranian approach*: The UAE and the US are both seeking to demonstrate to Middle Eastern audiences (including Iran's domestic ones) that Iran's claims of Western countries wishing to "deny Muslim states advanced nuclear technology" do not hold true. The message is also intended for the Iranian government: "it pays to play by our rules."[104]
3) *Champion of fast implementation*: Among the structurally very similar Gulf monarchies, competition for the positions of the first, biggest and fastest is rife. Abu Dhabi has chosen the fast track to nuclear power, that is, outsourcing the entire program instead of developing the needed domestic expertise. The Emirate is to gain a lot of international attention by successfully finishing the construction of the four first reactors in time—as is South Korea. In this process, good relations with key suppliers are paramount. A UAE official has described the country's approach to implementation as having "access to all of the major nuclear suppliers [in order to] ensure the long-term sustainability of the program."[105]
4) *Multilateral champion*: Championing engagement with key international actors and participating in UN agencies has been a further area of status-seeking. As the Minister of State for Foreign Affairs Anwar Gargash has said: "[the] UAE considers international partnership and cooperation as the cornerstone for the success of nuclear energy programs. This was a main pillar of the UAE Policy. We believe that the technologically advanced countries should make increasing efforts to facilitate the uses of peaceful nuclear energy to the less advanced countries."[106] Furthermore, the UAE has given "its share" to the support of a multilateral fuel bank initiative, in the form of US$10 million.[107]

[103] Hamad Al Kaabi, Permanent Representative of the UAE to the IAEA, quoted in Chris Stanton, "UAE seeks nuclear leadership role," *The National* (Abu Dhabi: September 17, 2010).

[104] See Jackson, "Nuclear Energy and Proliferation Risks", p. 1159.

[105] Anonymous UAE official quoted in Sean McLain, "UAE sets peaceful precedent in nuclear design", *The National* (Abu Dhabi: August 23, 2010).

[106] Anwar Gargash, Minister of State for Foreign Affairs, quoted in Emirates News Agency, "UAE's nuclear power to boost power grid", *Gulf News* (Dubai: September 22, 2010).

[107] Interview with high-level US diplomat, Helsinki (September 2008). The diplomat suggested the idea was his.

5) *Example of transparency, safety and security*: The local regulator, FANR, has lauded the program for upholding "the highest standards of safety, security, non-proliferation and operational transparency."[108] A prestige boost was offered by the nomination of the former IAEA chief Hans Blix to chair an international advisory board that will provide oversight for the UAE's nuclear program.[109]
6) *Champion of regional non-proliferation*: The UAE declares support for a weapons of mass destruction free zone (WMDFZ) in the Gulf. Prestige points were gathered with the UAE's election (instead of Iran's that was also running) to the board of governors of the IAEA for the term 2010–12, which was explained as reflecting "the care of [the] UAE's work for the IAEA."[110] Foreign Minister Sheikh Abdullah has also suggested that Abu Dhabi should be the host to an international conference to establish a WMDFZ in the Middle East.[111]
7) *Demonstrating the "peaceful intentions" of the UAE's foreign policy*: According to a government website, the guiding principles of the UAE's foreign policy are justice in international relations, including non-interference, the pursuit of peaceful resolutions of disputes, and support for international institutions, including the United Nations.[112] Vis-à-vis Iran, the UAE has asserted it "totally rejects the use of force as a solution... and rather calls for a solution through political means."[113] The preclusion of domestic enrichment and reprocessing can also be interpreted as demonstrating moral resolve to solve the Iranian question peacefully, in accordance with the country's foreign policy principles.

[108] Himendra Mohan Kumar, "UAE nuclear energy program 'very safe' *Gulf News* (Dubai: August 24, 2010).
[109] According to an official statement: "the IAB will provide the UAE nuclear program with the benefit of the expertise and knowledge of a highly select group of internationally recognised experts in the fields of peaceful nuclear energy". "Hans Blix to lead UAE nuclear panel", *UAE Interact* (February 23, 2010).
[110] Hamad Al Kaabi (Permanent Representative of the UAE to the IAEA), quoted in Emirates News Agency, "UAE enthusiastic partner of the IAEA initiatives", *Khaleej Times* (Dubai: September 29, 2010).
[111] Statement by Sheikh Abdullah bin Zayed Al Nahyan, the General Debate of the 65th Session of the United Nations General Assembly.
[112] "Foreign Policy", *UAE Interact* at http://www.uaeinteract.com/government/foreign_policy.asp (last accessed November 25, 2010).
[113] Tareq al-Haidan, UAE Assistant Foreign Minister for Political Affairs, quoted in Cynthia Johnston, "Analysis—UAE lining up with West over a nuclear Iran", *Reuters* (July 23, 2010).

Conclusions and future considerations

From Abu Dhabi's point of view, the fast development of the UAE's nuclear program can so far be deemed a success. With its publicity campaign regarding the motivations and justifications of its nuclear program the Emirate's elite managed to convince key international audiences of the country's motives. As a result, major external players not only approved the Emirate's plans but competed fiercely for participation in the project. While the main motivation behind the nuclear program was without doubt the looming domestic energy security crisis, and while prestige-seeking has become a central aspect of the UAE's nuclear diplomacy, as explained above, important challenges and problems relating to the "UAE model" have arisen en route.

The first set of challenges relates to domestic energy security. Firstly, the four reactors the Korean consortium has agreed to build in Abu Dhabi have an extremely ambitious schedule, given that in 2008 the federation still not did have any of the needed legal, institutional or infrastructural frameworks in place. The plan is to have the reactors operating in 2017–2020, one each year. According to a UAE-based nuclear regulator's estimate, "everything needs to work for the schedule to hold."[114] The chair of the International Advisory Board of the UAE's nuclear program, Hans Blix, has said the speed presents a "considerable challenge."[115] Secondly, even if Abu Dhabi were to beat all records and finish the four reactors by 2020, a gap between supply and demand might still emerge meanwhile, and the UAE would have to resort to either expensive LNG imports or dirty oil and coal. Solar energy would have to become considerably cheaper and fast so as to save Abu Dhabi's future energy leader the embarrassment of a return to hydrocarbons.

Secondly, outsourcing the entire program may indeed be the best option for a country lacking domestic know-how that chooses to prioritize rapid implementation and quick availability. However, it also creates important long-term dependencies on skilled foreign workforce. Due to the recent and fast socioeconomic development and small national workforce, the UAE, similarly to the other GCC states, has already for decades relied on imported expertise in key sectors, including energy, security, finance and strategic planning. In this sense, the nuclear energy program is no exception. Reflecting the

114 Interview with Christer Viktorsson, FANR, Abu Dhabi (May 2010), conducted by author.

115 Morris Loveday, "UAE's nuclear schedule poses "considerable challenge", says Blix," *The National* (Abu Dhabi: April 4, 2010).

magnitude of the endeavor, the establishing phase will, according to FANR, bring to the country around 10,000 construction workers and a couple of thousand skilled employees. The operation phase, estimated to require around 500 people,[116] however presents the most challenges in terms of securing this expertise in the country. With four reactors of a designed lifetime of 60 years,[117] the most sustainable strategy is to seek to balance the dependence on contractor services with the training of Emiratis. This will be pursued through the establishment of the Gulf Nuclear Energy Infrastructure Institute at Khalifa University of Abu Dhabi with the assistance of Texas A&M University and the US government-owned Sandia National Laboratories.[118] This arrangement will hopefully speed up employment of Emiratis in the program and also allow the US to closely engage with the UAE's future nuclear experts, possibly decreasing the probability of future proliferation. Nevertheless, it is doubtful whether Emiratis will ever make up a majority in the staff at ENEC and FANR. Since this is the first time in history that a nuclear program is based almost completely on outsourced services, this structural dependence adds an important long-term uncertainty factor and a management challenge for Abu Dhabi's government.

Thirdly, although dubbed internationally as "the UAE's nuclear energy program," the endeavor is tightly under the control of Abu Dhabi's ruling elite. The magnitude of the project and its implications for the whole country's energy security are likely to have an impact on the UAE's federal structures and dynamics, further increasing Abu Dhabi's already growing domination of the country's affairs in both domestic and international spheres. The UAE's constitution grants each emirate sovereignty over its natural resources, and the energy scarce Northern Emirates have long been dependent on budget allocations and subsidized energy from Abu Dhabi, which owns close to 95 percent of the country's fossil fuel resources. Even Sharjah and Dubai have during the past decade become increasingly dependent on energy from Abu Dhabi and Qatar (via the Dolphin pipeline). With the actual cost of nuclear electricity in Abu Dhabi yet to be determined, it is also unclear on what terms and for what price it will provide power to its sister emirates that are the home to a large share of the Emirati population. At the time of writing, in the spring of 2011,

116 Interview with Christer Viktorsson conducted by author.

117 J.Q. Marques, "Evolution of Nuclear Fission Reactors: Third Generation and Beyond", *Energy Conversion and Management,* Vol. 51 (2010): p. 1775.

118 *AMEinfo*, "Khalifa University signs MoU to develop and operate Gulf Nuclear Energy Infrastructure Institute", press release (February 23, 2011).

with increasingly vocal calls from many UAE nationals for a more equal wealth distribution, it seemed likely that Abu Dhabi would have to balance the growing welfare disparities in one way or another, whether through budget allocations or by ensuring that the other emirates receive a "fair share" of the atom.

As for the UAE's prestige-seeking, there are at least five potential problems: authoritarianism, Dubai's past, tensions with neighboring countries, the preferential treatment loophole, and the uncertainty over the adoption of the UAE's model. Firstly, the UAE's authoritarian political system places important constraints on the freedom of speech and rule of law. Domestic debate on nuclear energy has been non-existent and is likely to remain so. The torture incident from 2009 was a harsh reminder of some of the serious problems the state, which wishes to portray a modern and civilized image of itself, is facing regarding equality before the law. Secondly, Dubai's past as a transit point for illicit technology transfers by the A.Q. Khan network,[119] as well as its close trade links with Iran, place a shadow on the non-proliferation efforts of Abu Dhabi. Illicit trade in nuclear material, according to experts, is easier to disguise in a country with a nascent nuclear project.[120] Thirdly, tensions and competition with the UAE's two large neighbors, Saudi Arabia and Iran, mean that speculation on the UAE's motives will never be satisfied with the economic argument, even under complete transparency, due to the dual-use nature of nuclear technology and know-how. Recent overtures regarding the UAE's stance on a military attack to prevent Iran from acquiring a nuclear weapon have not helped in this sense: in July 2010, the UAE Ambassador to Washington announced the country would prefer a military attack to a nuclear-armed Iran.[121] According to leaked US documents, similar comments were purportedly made in July 2009 by Abu Dhabi's Crown Prince Sheikh Mohammad bin Zayed Al Nahyan, who had been growing "increasingly nervous" about Iran's activities since as early as 2005, going as far as requesting nuclear weapons from the US in 2007.[122] Fourthly, the 123

[119] Christopher Blanchard and Paul Kerr, "The United Arab Emirates Nuclear Program and Proposed U.S. Nuclear Cooperation" (Washington, DC: Congressional Research Service Report for Congress, 2009), p. 10.

[120] Mark Hibbs, Senior Associate, Carnegie Endowment quoted in McLain "UAE sets peaceful precedent in nuclear design."

[121] Youssef Al Otaiba, UAE Ambassador to the US, quoted in Johnston, "Analysis—UAE lining up with West over a nuclear Iran".

[122] "Embarrassed silence in Gulf over WikiLeaks," *Associated Press* (November 29, 2010); Samyuktha Krishnappa, " 'Increasingly nervous' about Iran, UAE wanted

agreement includes a minute that grants the UAE the right to demand renegotiation of the agreement, should better conditions be granted to any other state in the Middle East in the future.[123] UAE officials have downplayed the importance of this condition, whereas international experts have been more concerned on the future of the 123 agreement as a standard, should for example Jordan or Saudi Arabia be granted the right to domestic enrichment.[124] Finally, the most important weakness of the UAE model is that it is not necessarily one that other countries will want to implement. In 2008, immediate positive effects were reported in bilateral MoUs between the US and Bahrain and Saudi Arabia.[125] For the time being, however, Jordan and Saudi Arabia have both hinted at possibly wishing to maintain the right to enrichment in any agreement with the US.[126] Moreover, according to a regional expert, the UAE's easy renunciation of its rights under the NPT caused a lot of commotion in the developing countries' G77 group, as this was interpreted as breaking ranks with the group.[127] The prospect of countries flocking to implement the UAE's model is therefore not yet definite. Modified versions of it seem like a more likely course of events, given the different national interests and prestige-related perceptions of all the different developing countries.

Finally, on prestige-seeking: as many authors have warned, prestige should not become the sole end of foreign policy. Dore has pointed out that "only nations susceptible to the attractions of prestige-as-an-end-in-itself are likely to be constrained to obey whatever norms of international conduct the subter-

---

'Predator' from US: WikiLeaks", *International Business Times* (November 29, 2010).

[123] "Equal Terms and Conditions for Cooperation" in "Agreement for Cooperation between the Government of the United States of America and the Government of the United Arab Emirates Concerning Peaceful Uses of Nuclear Energy" (December 29, 2009).

[124] Hibbs in McLain, "UAE sets peaceful precedent in nuclear design."

[125] Mark Fitzpatrick, Senior Fellow, IISS, quoted in ibid.

[126] See Sharon Squassoni and Kee Hon Chung, "Nuclear Choices in the Middle East", *Nuclear in Focus Blog*, Center for Strategic and International Studies (October 14, 2010) at http://csis.org/blog/nuclear-choices-middle-east; Amena Bakr, "Saudi may enrich uranium for nuclear power plants", *Arabian Business* (from *Reuters*, June 16, 2010).

[127] Meeting of the working group on *The Nuclear Question in the Middle East*.

ranean movements of world opinion may gradually bring forth..."[128] Indeed, the UAE is not a "foolhardy egocentric."[129] It clearly has had other good reasons for not seeking to maintain the right to enrichment, including the imperative of the domestic energy crisis, the US nuclear umbrella, and the domestic costliness of enrichment. Nevertheless, becoming an example of the utmost application of this norm has clearly had more to do with United States foreign policy mixed with the UAE's craving for international attention and reputation. As Wylie has noted, for weaker states, international opinion is often the most important measure of worth, self-image and potential influence.[130] In the case of Abu Dhabi and the UAE, this measure, vis-à-vis the West, has proven to be particularly sensitive.

[128] Dore, "The Prestige Factor in International Affairs," p. 207.

[129] Wylie, "Valuing Reputation and Prestige", p. 124, in reference to Morgenthau's characterization of leaders who pursue prestige for its own sake.

[130] Wylie, "Seeking Prestige," p. 31.

7

Lessons from Turkey's Long Quest for Nuclear Power

Mustafa Kibaroglu

The nuclear disaster experienced in Japan's Fukushima Daiichi nuclear power plant in March 2011 has caused most, if not all, states that have or aspire to have nuclear reactors to seriously reconsider this potent capability.[1] Yet, it seems unlikely that nuclear technology will lose its status within international political and scientific circles as both a major source of energy and as a symbol of geopolitical standing. Prior to the devastating earthquake and tsunami resulting in the nuclear disaster in northeastern Japan, the effects of which are yet to be seen, nuclear energy had taken its place on the short list of investment plans by Middle Eastern states, especially those of the Persian Gulf which are already energy-rich. It is not clear whether the Fukushima crisis will

[1] This paper draws, in parts, on two sources, one of which is author's own publication, "Turkey's Quest for Peaceful Nuclear Power," *Nonproliferation Review* Vol. 4, No. 3, (Spring-Summer 1997): pp. 33–44; and the other of which is a PhD dissertation by Ms Sebnem Udum, conducted under the supervision of the author, "Understanding the Nuclear Energy Debate in Turkey: Internal and External Contexts," (Bilkent University, Ankara, June 2010). The author would like to express his appreciation for the contribution of his doctoral student to parts of this paper.

dissuade these governments from going ahead with their long-term plans to build nuclear power plants or whether they will downsize their projects in the face of the constant possibility of serious nuclear accidents. What is certain, however, is the fear of proliferation felt by Western states whose immediate concern is to strike a balance between expected cash revenues from nuclear technology transfers and the political-military implications of the Gulf's large-scale investments in such sensitive technologies. Hence, Middle Eastern countries aspiring to nuclear technology may have to satisfy the expectations of Western suppliers not only in terms of large sums of cash, which they appear prepared to pay, but also in terms of providing the assurances necessary for mitigating fears of further nuclear proliferation regionally. At this critical juncture, Turkey's longstanding quest for nuclear power may provide useful lessons for Middle Eastern countries seemingly displaying a genuine interest in acquiring state-of-the-art nuclear technology.

Nearly half a century ago, Turkey launched the first feasibility studies for a nuclear power plant with a view to benefitting from the peaceful exploitation of nuclear energy. However, those attempts, and subsequent ones, did not come to fruition. Among a variety of reasons, extending from a lack of a well-defined national strategy to domestic political disturbances, the most significant was the West's fear of a re-transfer of nuclear materials and technology to a third party, that being Pakistan in particular. Moreover, India and Greece further fueled these fears by disseminating rumors regarding such a Pakistani-Turkish alliance. This reportedly led to pressure by the United States on supplier countries and firms to deny transfers of nuclear reactors and related technology to Turkey.

Notwithstanding the allegations of a connection with Pakistan, Turkey has become party to the Treaty on the Non-Proliferation Treaty on nuclear weapons (NPT) after signing it on January 29, 1969, and then ratifying it on April 17, 1980. Turkey also concluded a "full-scope" Safeguards Agreement with the International Atomic Energy Agency (IAEA) in 1982. Eventually, Turkey joined the other international nuclear non-proliferation efforts such as the Zangger Committee and the Nuclear Suppliers Group (NSG) in 2000, and moved toward signing and ratifying the Comprehensive Nuclear Test Ban Treaty (CTBT) in 2001. Additionally, true to the fundamental thrust of its foreign and security policy,[2] Turkey has also become party to international

[2] The guiding principle of Turkish foreign and security policy as laid down by Mustafa Kemal Ataturk, the founder and the first President of the Republic of Turkey, is "peace at home and peace in the world."

agreements that sought to prevent the proliferation of weapons of mass destruction, such as the Biological Weapons Convention of 1972 and the Chemical Weapons Convention of 1993. At the Extension and Review Conference of the NPT held in New York in April/May 1995, Turkey gave its full support to the "indefinite and unconditional extension" of the Treaty. Turkey also used its influence on the Turkic republics of Central Asia and the Caucasus, who had recently proclaimed their independence from the Soviet Union, encouraging them to behave the same way.

Turkey also assumed a full member status in the Conference on Disarmament (CD) in Geneva in 1997 after a long period of attending the meetings with merely observer status. Moreover, as a country that never sought to acquire weapons of mass destruction in the first place, Turkey endorsed efforts to strengthen the nuclear non-proliferation regime and the verification mechanism of the IAEA. Turkey paid much attention to the proceedings of the "Programme 93+2" study as an attempt to make IAEA safeguards inspections more comprehensive. This culminated in the Additional Protocol in 1998, which Turkey signed and ratified in 2001. Above all, Turkey has been a "staunch ally" of the North Atlantic Treaty Organization (NATO) countries since 1952, and because of its position directly adjacent to the Soviet Union, it carried a greater portion of the burden of defending the West against the East.[3] With such an outstanding record of observing its commitments to the non-proliferation regimes, the opposition posed by the United States, Canada and Germany has long been effective in impeding Turkey's attempts to benefit from peaceful uses of nuclear energy. It goes without saying that such attempts by Turkey were its unequivocal right under Article IV of the NPT.

Turkey is a rapidly growing country whose energy needs at present already extend well beyond its installed power-generating capacity, and forecasts for the future paint a similar picture. Despite outstanding success in completing power-generation projects in the 1970s and 1980s, Turkey suffered from frequent power outages in the 1990s that caused serious damage *inter alia* to its industrial output. In the 1990s Turkey could not attain a similar pace of financing dozens of projects of various sizes that would further exploit its power-generating potential. Even if Turkey could have done so, it would have only postponed the anticipated crisis unless it resorted to other energy

[3] For a comprehensive discussion on these matters, see Mustafa Kibaroglu, "Turkey," in Harald Müller (ed.), *Europe and Nuclear Disarmament*, Peace Research Institute Frankfurt (PRIF) (Brussels: European Interuniversity Press, 1998), pp. 16–93.

sources. Turkey's hydropower and thermal energy sources would not have been sufficient to meet the steady increase in its energy demands in the decades to come even if all were applied toward power generation. As of the year 2010, Turkey had a population of approximately 75 million, which is estimated to reach 100 million in 2030.[4] Turkey does not have significant reserves of oil or natural gas; therefore, it is dependent on foreign nations for power generation. Turkey has other sources of energy such as fresh water, coal, geothermal, wind and solar, as well as considerable reserves of uranium and thorium. Of all these, however, only water and coal have been properly exploited over the last decades to generate electricity. Overall installed capacity in 2008 reached 42 electrical gigawatts (GWe), of which 14 GWe came from hydroelectric, and another 28 GWe from thermal plants fueled with both domestic and imported coal, natural gas, and oil. In terms of electricity production, the total capacity reached 262 Terawatt hours (TWh) (hydro 53 TWh; thermal 307 TWh) with a net production of 194 TWh (hydro 36 TWh; thermal 156 TWh).[5] Turkey's main domestic energy sources, water and coal, have potentials of 125 TWh per year and 105 TWh per year, respectively.[6] Hence, these figures indicate that even if all major domestic sources are somehow put into energy production, demand will still exceed domestic supply. Consequently, the peaceful exploitation of nuclear power came to the fore once again as a strategy for Turkey to diversify its primary sources of energy.

This chapter aims to give an account of Turkey's serious efforts to benefit from peaceful uses of nuclear energy over the last five decades and to discuss how the skepticism of Western supplier countries (arising from the fear of a connection to Pakistan) has had a negative effect on Turkey's initiatives in the past. After nearly five decades of sustained efforts to conclude deals with nuclear technology suppliers in the West, Turkey finally signed an intergovernmental agreement with Russia in May 2010 for the construction of its first nuclear power plant in southern Turkey. The deal, however, was considered by

[4] The population growth rate in Turkey is 2.1 percent per annum.

[5] The average productivity of hydroelectric plants is 70 percent, while productivity of plants fueled with coal, oil and natural gas is 75 percent on average. Gross electricity production was 8.6 TWh in 1970; 25 TWh in 1980; and 57 TWh in 1990. Figures sourced from State Hydraulic Works (DSI) of Turkey at http://www.dsi.gov.tr/faaliyet_raporlari/2009_faaliyet_raporu.pdf.

[6] Lecture notes of Ozden Bilen, former Director of State Hydraulic Works of Turkey, Hydropolitics and Strategical Research and Development Centre, Hacettepe University, Ankara (Spring 1996).

many analysts in Turkish scientific and technical circles as falling short of satisfying Turkey's initial aspirations to develop indigenous scientific and technological advancement in the nuclear field.

Turkey's past attempts to exploit nuclear energy

Initiation of peaceful nuclear energy in Turkey, albeit at the conceptual level, has its roots in Eisenhower's "atoms for peace" speech before the UN General Assembly in 1953, and the Geneva Conference of 1955 gathered in its aftermath. During the conferences, many of the secrets in the nuclear field were disclosed by the United States and the Soviet Union with a view to letting more countries make use of nuclear energy for peaceful purposes. The Turkish Atomic Energy Commission (TAEC) was then established in 1956 under the auspices of the Prime Ministry in order to coordinate efforts to establish nuclear research and training centers, as well as to issue licenses for nuclear power plants.

First phase: The 1960s and 1970s

Turkey's first nuclear research and training center, the Cekmece Nuclear Research and Training Center (CNRTC), was established in 1961. That was followed by the installation of a 1 thermal megawatt (MWth) pool-type research reactor TR-1 in CNRTC a year later.[7] Then, in 1966, the Ankara Nuclear Research and Training Center (ANRTC) was established in the environs of the capital, as the second major branch of the TAEC, for carrying out "fundamental and applied research to use nuclear energy and technology for the benefit of the country and to support the national development." With a view to exploiting natural uranium reserves of Turkey, the first feasibility studies for the construction of a 300–400 electrical megawatts (MWe) pressurized heavy water reactor (PHWR) were launched in 1967 so as to start generating electricity by the year 1977. However, domestic economic and political developments halted that initiative. The intervention of the military in domestic politics, on March 12, 1971 with the so-called "generals' memorandum" resulted in drastic changes in the government, which was administered for a

[7] CNRTC was founded on a 3,200-acres area beside the Kucuk Cekmece lake on the outskirts of Istanbul. Nuclear research and training are being carried out in its ten departments. See the website of the Turkish Atomic Energy Authority (TAEK) at http://www.nukleer.gov.tr.

couple of years by a series of short-lived ministerial cabinets. In a second attempt, the Nuclear Power Plants Division of the Turkish Electricity Authority (TEK), which had undergone reorganization in 1970, carried out comprehensive feasibility, site selection, and bid specification studies between 1972 and 1974 for a 600 MWe nuclear power plant.[8] Surveys for the selection of a nuclear site were conducted throughout Turkey. Due to its stable seismic conditions, Akkuyu Bay on the Mediterranean coast about 43 k.m. southwest of Silifke, was selected. Then, TAEC issued a license in 1976 for the site selected by TEK. A consortium of one French and three Swiss firms assisted in preparing the necessary paperwork for bidding. Negotiations on the construction of a 600 MWe nuclear power plant, fuel services, and the financing of the investment were begun in 1977 with two half-state-owned Swedish firms, Asea-Atom and Stal-Laval, which had made the best offer.[9] These negotiations were interrupted by the Turkish military coup in 1980.[10]

Second phase: The 1980s

In 1979 the 250 thermal kilowatts (KWth) Triga Mark II research reactor started up operations and in 1981 the TR-1 research reactor, which was shut down in 1977, was replaced by a 5 MWth pool-type research reactor TR-2. In early 1980, a site selection survey for a second nuclear power plant had already started. Then, the site was selected in the northernmost point of Turkey in the central Black Sea region at Inceburun, located some 25 k.m. west to Sinop.

Notwithstanding these renewed initiatives by Turkey, the early 1980s also marked the beginning of allegations of illicit cooperation between Turkey and Pakistan. In 1981, the United States expressed concerns about a Turkish-

[8] Kenan Unlu, "Turks Take Steps to Revive Their Nuclear Programme," *Nuclear Engineering International* (Kent: January 1995), pp. 16–17.

[9] See Ahmet Kutukcuoglu (former Director of Nuclear Power Plants Division of TEK), "Turkiye'nin Gecmisteki Nukleer Enerji Deneyimleri" [Past Experiences of Turkey in the Nuclear Field], *Uluslararasi Nukleer Teknoloji Kurultayi* [International Nuclear Technology Forum] [October 12–15, 1993], Ankara Makine Muhendisleri Odasi [Chamber of Mechanical Engineers, Ankara], No. 168 (March 1994): pp. 40–4.

[10] The third of the so called *once-in-a-ten-year* intermissions in Turkish democracy took place on September 12, 1980 (previously on May 27, 1960 and March 12, 1971) as a result of the military coup of the then Joint Chiefs of Staff. The democratic regime was reestablished with the country-wide elections held in November 1983.

Pakistani alliance on the grounds of alleged shipments from Turkey to Pakistan of strategic material with potential nuclear weapons implications, which would enable Pakistan to go ahead with its quest for uranium enrichment technology. It was reported that US Ambassador in Ankara Robert Strausz-Hupe discussed the matter with Turkish Minister of Foreign Affairs Ilter Turkmen. On this particular account, Ambassador Omer Ersun, who was then Chief of the Policy Planning Staff at the Turkish Ministry of Foreign Affairs, confirmed that the United States administration raised concerns about the US$30,000 shipment of inverters to Pakistan by a Turkish textile firm. Ambassador Ersun added that neither the Turkish Ministry of Foreign Affairs nor the private textile manufacturer had any specific knowledge that the inverters would be intended for Pakistan's uranium enrichment program.[11]

Allegations of an illicit Turkish-Pakistani relationship indeed owed much to the unique character of relations between the two countries. For almost any Turkish man in the street, Pakistan was long believed to be the one and only country really friendly to Turkey.[12] There is a great sympathy in the public domain of both countries. In fact, the revolutionary thoughts and the principles of Mustafa Kemal Atatürk inspired the Pakistani people in their fight for independence against the British. As an indication of this, Atatürk's name has been given to a variety of institutes, libraries, and to beautiful districts in big cities of Pakistan.[13] Turkey's historic relations with Pakistan gained momentum with the Agreement of Regional Cooperation for Development (RCD) of 1964 which brought together Turkey, Iran and Pakistan, in fact as another ring in the chain of the US "containment" strategy against the Soviet Union.[14]

[11] Telephone conversation with Omer Ersun, Turkish Ambassador, Ottawa (March 24, 1997).

[12] A very common belief in the public domain in Turkey can be translated as "there are no friends of Turks but the Turks," which is quickly followed up with "except the Pakistanis."

[13] The same applies to Pakistani leaders such as Mohammed Ali Jinnah and Zulfikar Ali Bhutto, whose names are given to institutes, libraries and districts in Ankara or Istanbul. As a very recent confirmation of this observation, the speech of Humayun Han Banghash, the Pakistani Ambassador to Ankara, during a reception to celebrate the 50th anniversary of the foundation of Pakistan, is remarkable. Ambassador Banghash said, "the Pakistani people have once more demonstrated how much they admire Atatürk [by hanging everywhere posters of Atatürk and Jinnah]," *Hurriyet* (Istanbul: March 21, 1997), p. 1.

[14] The RCD remained in existence until 1979. In 1985 the RCD, by then dormant, was resuscitated as the Economic Cooperation Organization (ECO). After the

Friendly relations were further intensified in many respects in the aftermath of the military coup in Turkey on September 12, 1980. Military leaders of Turkey and Pakistan, namely President General Kenan Evren and President General Zia ul-Haq, paid a series of reciprocal visits to each other's countries until the latter was killed in a plane crash on August 17, 1988. These ceremonial visits increased the magnitude of trade and cooperation in many fields, including civilian and military spheres.

In 1982, the TAEC was abolished and succeeded by the Turkish Atomic Energy Authority (TAEK) which was established by law as a governmental organization under the direct supervision of the prime minister.[15] The main duties of TAEK, as stated in related Act No. 2690, can be summarized as: "to determine and progress the basis of the national policy and the related plans and programs and to submit them to be approved by the prime minister; to execute and to support scientific, technical and administrative studies; to give approval, permission and license to the nuclear installations; to enlighten the public in nuclear matters; to establish research and educate the personnel in the nuclear field."[16]

In order to reinvigorate the stalled efforts, seven major suppliers were invited in 1983 to submit bids to install nuclear power plants in Turkey. Eventually, letters of intent for the supply of three power reactors in two sites were issued to three firms, namely Atomic Energy of Canada, Ltd (AECL) for a 655 MWe CANDU reactor in Akkuyu; Kraftwerk Union (KWU) of Germany for a 990 MWe PWR in Akkuyu; and General Electric (GE) of the United States for one or two 1,185 MWe BWR in Sinop. The total cost of these three units was estimated to be US$3.4 billion.[17]

---

disintegration of the Soviet Union, Turkey took an initiative to enlarge the Organization to include Azerbaijan, the Central Asian Republics and Afghanistan. See the website of Turkish Ministry of Foreign Affairs at http://www.mfa.gov.tr.

[15] Yalcin Sanalan, "Turkiye Atom Enerjisi Kurumu'nun Nukleer Enerji Uretimindeki Yeri" [The Role of TAEA in the Development Nuclear Energy], *Uluslararasi Nukleer Teknoloji Kurultayi* [International Nuclear Technology Forum] [October 12–15, 1993], Ankara Makina Muhendisleri Odasi [Chamber of Mechanical Engineers, Ankara], No. 168 (March 1994): pp. 128–34.

[16] TAEK, "Law on Turkish Nuclear Atomic Energy Authority", No. 2690 at http://www.taek.gov.tr/eng/legislation/func-directinfo/587/.

[17] "The Chosen Three," *Nuclear Engineering International* (Kent: December 1983), p. 4.

Talks with Canadian and German firms, and allegations

Preliminary results of the site surveys conducted in Sinop by an expert team dispatched by General Electric, and that team's consultations with the experts from the IAEA and from Turkey who had previously carried out similar studies in the region, did not furnish positive results as to the feasibility of a nuclear power plant. Without conducting much more elaborate and comprehensive studies on the risk of an earthquake in the region, no further steps could be taken for the construction of a nuclear site. Hence, negotiations with General Electric came to an end. However, negotiations with AECL and KWU were pursued throughout 1984 and the parties agreed to the preparation of contract documents and financing schemes for a turnkey approach. Right at that juncture, the newly established government in Turkey suggested a so-called "build-operate-transfer" (BOT) model for the supply of power reactors. The BOT model was launched with a view to attract foreign investments to Turkey in many fields, and the nuclear deal was also considered in the same vein. With BOT, the Turkish government suggested AECL and KWU should construct the nuclear sites by forming joint venture utilities (JVU) with Turkey's TEK; then operate the reactors for 15 years by selling the generated electricity to TEK; and then to hand over their share in the nuclear site to TEK. AECL and KWU were also both required to be responsible for a greater portion of financing the power plants that they would build in the form of JVU with TEK.[18] While both firms in principle agreed to the scheme, KWU later declined to undertake the supply of the reactor—due, it was reported, to a disagreement over the conditions of financing imposed by the Turkish government.

Hence, AECL was left as the sole potential supplier for all three reactors. On January 3, 1985, Turkey invited AECL to conduct final negotiations for the supply of reactors for its Akkuyu and Sinop nuclear units.[19] The Canadian firm was said to have made serious attempts to find loan guarantees to finance the Akkuyu reactor and requested a US$1 billion loan guarantee both from the Canadian government and from the banks.[20] However, negotiations with

[18] For a precise account on this matter, see Ahmet Kutukcuoglu, "Turkey's Joint Venture Scheme," *Nuclear Engineering International* (Kent: March 1986), p. 29.

[19] "Turkey: AECL Close, KWU Still Running, W-MHI Out," *Nuclear News*, (February 1985), p. 102.

[20] Diane Francis, "Bank risk share asked in Turkish reactor deal," *The Toronto Star* (Toronto: November 13, 1985), p. G1. Sourced from CNS Nuclear Databases.

AECL ran into problems upon the failure of the Canadian government to approve the plan.[21] Canada requested official guarantees from the Turkish state regarding finances, and submitted a proposal containing a credit package and a number of conditions. However, the proposal was not found by the Turkish government to be consistent with the requirements of the BOT scheme, and the deal was reported to be suspended.[22] An AECL official, however, maintained that they were still negotiating with Turkey for a contract to build and operate a CANDU reactor near Akkuyu.[23] Meanwhile, another AECL official was quoted as admitting that "it [was] not possible for an agreement to be reached within the framework of a build-own-run formula with the Turkish government."[24] Yet, in January 1987, Prime Minister Turgut Ozal was quoted as saying that Turkey was ready to evaluate further proposals from AECL to go ahead with the reactor deal.[25] And, in June 1987, it was reported that negotiations with AECL continued after the Canadian government failed to approve a plan for AECL to build the plant and run it for 15 years. It was also reported that talks with KWU were reopened.[26]

Nevertheless, none of these talks have come to fruition, not only because of financial problems, but also because of the political considerations of the Canadian and German governments. The withdrawal of KWU from the deal is said to have arisen mainly from the then West German government's reaction to Turkey's improving relations with East Germany. Although such a view cannot be attributed to any official communication or document, it was so perceived in the Turkish diplomatic circles. With regard to the Canadian firm, on the other hand, it was reported that AECL had withdrawn its bid for a nuclear plant "in response to pressure from Western countries which [were] concerned that Turkey may build a nuclear bomb based on CANDU

[21] "Candu Sales Hit Problems," *Nuclear Engineering International* (Kent: October 1986), p. 14.

[22] Unlu, "Turks Take Steps to Revive their Nuclear Programme," p. 16.

[23] "AECL Says That it is Still Negotiating with Turkey," *Nuclear News* (November 1986), p. 21. AECL's statement was made in response to a report by *Reuters* (October 15, 1986) in which the Turkish Minister was referred to as saying the deal was suspended.

[24] Ray Silver, "Atomic Energy of Canada, Ltd. (AECL) is not Conceding Either Turkish or Argentinean Prospects for CANDU Sales," *Nucleonics Week* (January 22, 1987), pp. 8–9.

[25] Ibid.

[26] "Still Negotiating," *Nuclear Engineering International* (Kent: June 1987), p. 38.

technology."[27] Opposition from Greece, Israel and France resulted in a deadlock in efforts to procure financing for the project.[28] Western countries feared that Turkey would do what Pakistan did—modify the technology to gain the capability for building an atomic bomb.[29] When allegations began to heighten, especially following an Indian journalist's interview with Abdul Qadeer Khan, a key figure in Pakistan's nuclear program, Ambassador Inal Batu, then spokesperson of the Turkish Ministry of Foreign Affairs, explicitly denied reports that Turkey would sell material for nuclear arms production to Pakistan, in a press conference held on October 28, 1987. Ambassador Batu said that "Turkey fulfilled with great care its obligations under the NPT."[30] Later, Indian sources reported that the Indian government was looking over reports that Turkey was illegally selling nuclear weapons-related materials to Pakistan. Kapil Verma, from the Indian Congress Party, was quoted as saying that "India should try to persuade the Turkish government to give up their strategic nuclear equipment."[31] Again in 1988, Leonard S. Spector, a US expert from the Carnegie Endowment for International Peace, indicated that "although the United States had put an end to the smuggling of sensitive material by Pakistan, both India and Pakistan had accumulated the nuclear materials necessary for bomb production."[32] Building upon Leonard Spector's views, Indian authorities argued that "Turkey was used as an intermediary in many of Pakistan's clandestine activities" even though Spector did not cite any particular country by name.[33]

Talks with Argentine firms, and allegations

When talks with Canadian and German firms for the construction of nuclear power plants ran into trouble, Turkey sought other partners in the nuclear field so as to acquire nuclear technology and material. Hence, Turkey signed

[27] "Canadian Firm Drops Bid to Build Nuclear Plant," *Nuclear Developments* (February 25, 1988), p. 39. Sourced from CNS Nuclear Databases.

[28] Ibid.

[29] Ibid.

[30] "Spokesman Denies Nuclear Sales to Pakistan," *Nuclear Developments* (January 28, 1988), p. 95. Sourced from CNS Nuclear Databases.

[31] "MP Notes Turkish Nuclear Equipment to Pakistan," *Nuclear Developments*, April 1, 1988, p. 5. Sourced from CNS Nuclear Databases.

[32] "U.S. Expert Interviewed in South Asian Nuclear Arming," *Nuclear Developments*, September 2, 1988, pp. 17–19. Sourced from CNS Nuclear Databases.

[33] Ibid.

a fifteen-year nuclear cooperation agreement with Argentina on May 3, 1988. The agreement was said to pave the way to potential transfers from Argentina including technical assistance for front-end nuclear fuel cycle research and development, power and research reactor planning, construction, quality assurance, operation and regulatory research. The IAEA safeguards would apply to all nuclear material designed for the use of nuclear technology to be transferred under the accord. It was also agreed that safeguards on Argentinian supplies would continue even if the agreement expired. In the same context, it was repórted that Turkey had shown interest in the 380 MWe Argos PWR design unveiled a year before by Empresa Nuclear Argentina de Centrales Eléctricas (ENACE).[34] Argentina would also help Turkey study the feasibility of a site for Argos in Turkey. As a consequence of this agreement, Turkish scientists and technicians would go to Argentina's Bariloche Nuclear Center for training. Other areas for cooperation between the two countries were also explored, including uranium mining, nuclear fuel plants, industrial production of radioisotopes, and safety requirements.[35] It was also reported that Turkey was seeking a joint agreement with Argentina to participate in the construction of an Argentine produced 25 MWe nuclear reactor, namely CAREM-25, in Ankara. Prof. Dr Atilla Ozmen, then Director of TAEK, was quoted as saying that "the offer [would] include financial help, as well as technical personnel."[36] The CAREM-25 reactor was developed by Investigaciones Aplicadas (INVAP) at Bariloche, to be fueled with 4 percent enriched uranium. During a visit in late April 1989, Prof. Ozmen said that "Prime Minister Ozal [was] looking to follow the example of Argentina in achieving independence in certain high technology fields such as nuclear energy."[37] He said, "Turkish engineers, scientists and companies would fully participate in construction of Argentine reactor, and the eventual purchase of Argentine [Argos] reactor [would] also help ease power shortages in Turkey forecast for the mid 1990s."[38]

[34] ENACA is 75 percent owned by Argentine Comision Nacional De Energia Atómica (CNEA) and 25 percent owned by KWU of Germany. Richard Kessler, "Argentina Says Nuclear Accord With Turkey Sets Stage for Exports," *Nucleonics Week* (May 12, 1988), pp. 8–9.

[35] "Argentina and U.S. in Co-operation Talks," *Nuclear Engineering International* (Kent: November 1988), p. 3.

[36] "More Argentine Sales?", *Nuclear Engineering International* (Kent: July 1989), p. 5.

[37] Richard Kessler, "Argentina Says Nuclear Accord with Turkey Sets Stage for Exports," *Nucleonics Week* (May 12, 1988), pp. 8–9.

[38] Ibid.

Argentine officials were hopeful Turkey would buy a 380 MWe Argos PHWR from vendor ENACE, provided the CAREM-25 reactor project worked out well.[39]

In October 1990, Turkey and Argentina agreed on forming a joint architectural-engineering firm to develop Argentina's modular low power CAREM-25 reactor. Two Turkish firms Sezai Turkes-Fevzi Akkaya (STFA) and TEK, and two Argentine firms namely, Comisión Nacional de Energia Atómica, and INVAP formed the new joint firm. The agreement included a commitment to build two 25 MWe CAREM-25 units, one in each country. While Turkey agreed to provide most of financing for the work, Argentina agreed to provide most of the technology. Accordingly, Argentina would provide the Nuclear Steam Supply System technology, basic and detailed engineering for the balance of plant, construction management, and regulatory expertise. If preparations for the building of two units would go ahead as planned, work on the first unit in Argentina would begin in 1991, and construction of the second unit, in Turkey, would begin in 1992.[40] The long-term goal of the joint venture was said to have been to export the reactor to other nations in Latin America, Africa and the Middle East.[41] Turkish Prime Minister Ozal and Argentine President Carlos Menem had corresponded and met regarding the project, and hence played a key role in obtaining the agreement.

Although elaborate and high level talks between Argentina and Turkey culminated in a formal document, no progress has been made since then. It was reported that Argentina's decision to appoint Adolfo Saracho, a former head of Argentina's nuclear energy commission, as its new Ambassador in Ankara made "the United States, the Soviet Union, and German diplomatic missions in Buenos Aires uneasy."[42] When Turkey and Argentina agreed in

[39] Ibid.

[40] "Agreement Signed to Build Carem," *Nuclear Engineering International* (Kent: December 1990), p. 8. Also see Richard Kessler, "Argentina and Turkey to Form Nuclear A/E to Build a Small PWRs," *Nucleonics Week* (October 25, 1990), pp. 12–3. For more on this issue see "Nuclear Pact With Argentina 'Secretly Signed,'" *Nuclear Developments* (November 15, 1990), pp. 29–30. Sourced from CNS Nuclear Databases.

[41] "Reactor Venture Entered With Turkey," *Nuclear News* (December 1990), p. 46.

[42] It was reported that Saracho made a significant contribution to the signing on May 3, 1988 of a nuclear technology cooperation agreement. Saracho was quoted as saying that "Argentina's objective [was] to help Turkey develop its nuclear technology without depending on any country or bloc, and hope[d] to help Turkey build reactors and nuclear power plants in the future." "Argentina to Help Acquire

October 1990 to set up a joint venture company to build CAREM-25 reactors in each country, "the United States, the Soviet Union, and Germany believed that Turkey's acquisition of nuclear technology would be disadvantageous."[43] Hence, the United States was said to have "worried because of the fears that Turkey might sell this technology to Pakistan."[44] William Rope, First Undersecretary at the US Embassy, was reported to have made several calls trying to determine how nuclear cooperation between Turkey and Argentina would affect other countries in the region, especially Pakistan.[45]

The CAREM-25 project was canceled a year later by the Turkish Atomic Energy Authority, in a unilateral decision of the executives. Prof. Yalcin Sanalan, the director of TAEK at the time, noted that "[he] found the prospects of the CAREM-25 deal ambiguous" on the grounds that "CAREM-25 was too small for electricity generation and too big for research or training, however, very suitable for plutonium production," which would make the project of proliferation a concern; therefore, Prof. Sanalan concluded that "such an ambiguous project would decrease the chances of Turkey in its current and future quest for large-scale nuclear power plants, which the country really needed."[46] Interesting to note, Prof. Sanalan also revealed that that "when TAEK declared that it unilaterally canceled the CAREM-25 project, as the director of TAEK [he] was then frequently invited to 'inner circle' meetings and dinners of the OECD/NEA"; and that "the OECD countries then openly expressed in these meetings their wishes to help Turkey acquire nuclear power plants, provided the latter had well defined objectives in that area."[47] A mere implication of the attitude of the OECD countries prior to and after the unilateral cancellation of the CAREM-25 project by Turkey points to their proliferation concerns over the acquisition of advanced nuclear technology and material by Turkey.

'Nuclear Technology'," *Nuclear Developments* (June 21, 1988), p. 39. Sourced from CNS Nuclear Databases.

[43] Ibid.

[44] "Nuclear Pact With Argentina 'Secretly Signed'," *Nuclear Developments* (November 15, 1990), pp. 29–30. Sourced from CNS Nuclear Databases.

[45] "'Secret Talks' With Argentina on Nuclear Plant," *Nuclear Developments* (October 6, 1989), pp. 31–2. Sourced from CNS Nuclear Databases.

[46] E-mail correspondence with Prof. Yalcin Sanalan, Department of Nuclear Engineering, Hacettepe University, Ankara (February–March 1997).

[47] Ibid,

Third phase: the 1990s

Turkish experts in the early 1990s made estimates for the scale of nuclear energy that would be needed for the decades to come. Figures indicated that in order to meet Turkey's ever-growing energy needs a 1,000 MWe per annum capacity would be needed by 2005. In the same vein, total installed nuclear power capacity should reach 34,000 MWe by 2040, an estimate based on both the population growth rate and the yearly increase in energy consumption in Turkey estimated for the first half of the 21st century. Hence, an approximate and rather stable 5,000 kilowatt hour (KWh) per capita consumption paired with a population of 110 million in the year 2040 results in an estimated demand of 550 TWh of energy for that period. Considering that a 1,000 MWe nuclear reactor capacity can generate 6,750 MWh electricity per annum, operating at 70 percent productivity on average, then 34,000 MWe of installed capacity is estimated to generate 230 TWh of energy in 2040.[48] Although the above scenario envisions a solution to Turkey's long-term energy needs, the installation of some thirty-five large-scale nuclear power plants seems hardly possible, due to both technical and financial constraints. The disintegration of the Soviet Union and Turkey's intensified relations with the Turkic republics in Central Asia and the Caucasus, some of which have nuclear installations, raised further concern for the West.

One of the most significant accusations regarding Turkey's alleged role in Pakistan's nuclear weapons program came from US Senator John Glenn. In his article published in the *Washington Post* on June 24, 1992, Senator Glenn pointed out to the history of the Reagan and Bush administrations in dealing with the issue of nuclear proliferation, especially in regards to Pakistan. Senators John Glenn and Stuart Symington had amended the Foreign Assistance Act in 1977 so that no country could receive US aid if it imported or exported unsafeguarded nuclear enrichment or reprocessing materials and equipment. Hence, Pakistan's aid was cut off in 1979 for violation. In the article, Senator Glenn claimed that "the credibility of the Bush and Reagan administrations' commitment to nuclear non-proliferation has been destroyed by their willful

[48] For details, see Dogan Oner (former Deputy Director of TAEA), "Turkiye'nin Nukleer Enerji Santralleri Konusunda Ileriye Donuk Politikalari Ne Olmalidir" [What Should be Turkey's Forward Looking Policies About Nuclear Power Plants?], *Karayollari Vakfi Dergisi* [Journal of Highway Foundation] (April 1991): p. 19.

misinterpretation of the U.S. laws."[49] He said "U.S. assistance allowed Pakistan to divert funds to its nuclear program, thus damaging the nuclear non-proliferation effort."[50] In that context, Senator Glenn also mentioned the "failure of the Glenn-Symington amendment to stop aid to Turkey, because of its involvement in aiding Pakistan in its acquisition of uranium enrichment equipment."[51] As a reaction to Senator Glenn's accusations, Turkish officials promptly denied that Turkey was supplying Pakistan with sensitive equipment used in the production of nuclear weapons. Recent accusations from US congressmen have also charged Turkey with helping Pakistan's nuclear program.[52]

Another source of concern about a possible Turkish-Pakistani nuclear relationship came from Greece. Dr Thanos Dokos of the Greek Defense Ministry stated that:

> [al]though Turkey does not possess the technological capability to develop nuclear weapons in the near future, Greece is worried of the alleged nuclear cooperation between Ankara and Islamabad, the rise of Islamic fundamentalism and (unconfirmed) reports that Turkey might try to acquire nuclear weapons material and technology and recruit nuclear scientists from the Muslim republics of the former Soviet Union.[53]

Various sources in Greece had already disseminated information to that effect. For instance, a Greek daily, *I Kathimerini*, alleged that "Turkey [was] attempting to acquire nuclear technology and nuclear warheads from the Islamic republics of the former Soviet Union, and [was] trying to acquire nuclear potential by cooperating with Pakistan in particular. The government of Greece [was] therefore deeply concerned by the development."[54] Ambassador Filiz Dincmen, spokeswoman of Turkish Ministry of Foreign Affairs,

[49] John Glenn, "This country encouraged the spread of nuclear weapons," *Washington Post* (June 24, 1992), p. A19.

[50] Ibid.

[51] Ibid.

[52] "Turkey: Reports of Nuclear Transfer to Pakistan Denied," *Proliferation Issues* (July 10, 1992), p. 29. Sourced from CNS Nuclear Databases.

[53] Unlike many other Greek-originating publications, Dr Dokos was kind enough to use words "alleged" and "unconfirmed" regarding a Turkish-Pakistani connection. But still, the fear of such a connection is obviously felt vividly in Athens. Thanos Dokos, "Greece," in Harald Muller (ed.) *Nuclear Export Controls in Europe* (Brussels: European Interuniversity Press, 1995), p. 208.

[54] *I Kathimerini* (Athens: December 29, 1991), p. 1. Cited in "Turkey Reportedly Seeking Soviet Nuclear Warheads," *Proliferation Issues* (January 31, 1992), p. 61. Sourced from CNS Nuclear Databases.

said that "no efforts had been made to obtain raw materials or technology used in the manufacture of nuclear weapons from the former Soviet republics of Central Asia," and recalled that "Turkey [was] a state party to the NPT and [had] full-scope safeguards agreement with the IAEA in force."[55]

Behind the allegations about a Turkish-Pakistani connection were those sources mostly from Greece and India; each was an adversary of one of the parties accused for being in illicit collaboration. Although the veracity and thus credibility of such allegations were highly suspicious, their probable effects on Turkey's failure to install nuclear power plants were worth considering seriously by Turkish administrations in order to avert another backlash in their successive initiatives to install such plants.

In early 1995, it was reported that construction tenders for a nuclear power plant at the Akkuyu site would be issued during the year. Later, the Turkish Electricity Generation and Transmission Company (TEAS) considered bids for consultancy contract for the power plant project. TEAS was reported to have received bids from at least 14 consortia including Turkish partner companies. In March 1995, the Korean Atomic Energy Research Institute (KAERI) with a consortium of Korean companies and Turkish engineering firms was said to be positioning itself as a contender in the upcoming bid to supply a nuclear plant to Turkey. KAERI had already begun work in Turkey in December 1994, along with the Turkish engineering firm GAMB.[56] KAERI then won the contract to evaluate the nuclear development program drawn up by TEAS. The contract's main objectives included: investigation of internationally accepted contemporary nuclear power plant types and systems feasible for Turkey; review of bids solicited during previous attempts to set up a nuclear program in Turkey; and consultation with TEAS during bid evaluations and contract negotiations.[57] KAERI examined the feasibility of renewing Turkey's project at Akkuyu. A review process was then scheduled for completion by mid-1996. A contractor would be selected by 1998, with construction scheduled to begin in late 1998. Atomic Energy of Canada was expected to offer a 680 MW CANDU-6 heavy water reactor, and Siemens of

[55] "Ministry Denies Trying to Get Nuclear Technology," *Proliferation Issues* (January 31, 1992), p. 67. Sourced from CNS Nuclear Databases.

[56] "Turkish Plant: Bids in 1995," *Nuclear Engineering International* (Kent: January 1995), pp. 6–7.

[57] TEAS said that KAERI's was the most comprehensive and, at US$350,000, the least costly of the bids it received. "Korea Prepares for Turkish Bid," *Nuclear Engineering International* (Kent: March 1995), p. 5.

Germany was said to offer a 1,400 MW pressurized water reactor.[58] According to the news in the Turkish media in December 1996, the Turkish government was planning to accept bids for the nuclear plant to be built either as single unit with a capacity of 1,200 MW or two equal units each with a capacity of 600 MW. The cost of the plant was estimated at about US$1.5 billion. However, because of the political developments in Turkey that resulted in the coming to power of a so-called "Islamist" government, bidders were said to be reluctant to go ahead with the nuclear power plant deal. On January 20, 1997, Turkey extended the deadline for bidding which expired that day at the request of the companies.[59]

Fourth phase: the 2000s

The Erbakan government stepped down following a political crisis in early 1997 and was succeeded by the Mesut Yilmaz government. Yilmaz was determined to strengthen relations with the United States. Then, the United States, Germany and France stepped up their lobbying activities for Westinghouse, the NPI (Nuclear Power International), and the Siemens-Framatome Consortium.[60] The Yilmaz government lasted until 1999. As a result of the 1999 general elections, a new coalition government was formed, headed by Bülent Ecevit as Prime Minister and including the Democratic Left Party (DSP), the Nationalist Action Party (MHP), and the Motherland Party (ANAP). The Energy Minister in the new coalition government, Cumhur Ersumer of ANAP, was determined on the establishment of the nuclear power plant in Akkuyu.[61] However, the Treasury did not give purchase guarantees, and this decision was supported also by the IMF and the World Bank.[62] The project ran into hurdles and was followed by corruption lawsuits.[63] The bidding com-

[58] Mark Hibbs, "Turkey Expected to Request Bids for PWR Project in Coming Weeks," *Nucleonics Week* (March 21, 1996), pp. 1–2.

[59] "Turkey Extends Nuclear Bid Enquiry Period," *Anatolia News Agency* (Ankara: January 21, 1997).

[60] Mark Hibbs, "World finance, regional gas market keys to Turkey's Akkuyu project", Nucleonics Week (August 28, 1997), pp. 10–3.

[61] "Ersumer'den Greenpeace'e Rest [Ersumer's Challenge to Greenpeace]," *Sabah* (October 21, 1999) at http://arsiv.sabah.com.tr/1999/10/21/p03.html.

[62] "Hazine Garantisizligine DB ve IMF Destegi [Support from WB and IMF to No-Guarantee [to the Nuclear Project] by the Treasury]," *Sabah* (April 23, 2000) at http://arsiv.sabah.com.tr/2000/04/23/e04.html.

[63] "Nukleer Santral Pazarligi Yapmis [Bargaining by Central Nuclear Authority],"

panies were in dispute;[64] the nuclear project competed with defense projects; and there were proliferation concerns.[65]

In April 2000, the Energy Minister, Ersumer, declared a decision by the government to postpone the nuclear plant project until July 2000 due to financial constraints as a result of the IMF-backed economic program.[66] On July 25, 2000, Prime Minister Ecevit declared that the tender for the nuclear power plants was cancelled due to the shortage of funds to construct the power plant.[67] The tender process led to several rows and rumors of corruption cases, and it was one of the reasons of the postponement by the government.[68] It was also argued that Ecevit canceled the tender because whichever firm won the contract for the Akkuyu power plant, the other two would sue the winning consortium for cancellation of the tender.[69] In the 1990s, Turkey faced financial problems for new projects on energy which were based on generating electricity from water and coal; therefore, it became necessary to seek alternative energy resources. Overall, the attempts failed because of the

---

Aksam (October 23, 2002) at http://www.aksam.com.tr/arsiv/aksam/2002/10/23/yazidizi/yazidizi1.html; "DGM'den Agir Ceza'ya [From the State Security Courts to Heavy Penalty Court]," *Sabah* (May 5, 2001) at http://arsiv.sabah.com.tr/2001/05/05/p09.html.

64 Ercan Ersoy, "Turkey nuclear bid may hit new challenges," *Reuters* (February 23, 2000).

65 Ezelhan Ustunkaya, "Hem Atomcu Hem Cevreci! [Both Atomist and Environmentalist!]," *Milliyet* (March 15, 2000) at http://www.milliyet.com.tr/2000/03/15/haber/hab09a.html; Zulfikar Dogan, "Nukleeri Karistiran Mektup," *Milliyet* (April 9, 2000) at http://www.milliyet.com.tr/2000/04/09/ekonomi/eko00.html.

66 "Nuclear-plant decision postponed until July," *The Globe and Mail* (April 22, 2000), p. A22.

67 "Ecevit: Akkuyu ile Ilgili Ihalenin Iptali Nukleer Enerjiden Vazgecmek Anlamina Gelmiyor [Ecevit: Cancellation of Akkuyu Bid Does not Mean Neglect of Nuclear Energy]," *BYEGM* (July 26, 2000) at http://www.byegm.gov.tr/YAYINLARIMIZ/HABERANADOLU/HABER-ANA/2000/07/HA00X07X26.HTM.

68 "Nukleer Santral Pazarligi Yapmis [[He] Negotiated on Nuclear Power Plant]," *Aksam* (October 23, 2002) at http://www.aksam.com.tr/arsiv/aksam/2002/10/23/yazidizi/yazidizi1.html; "DGM'den Agir Ceza'ya [From the State Security Court to Heavy Penalty [Court]]" *Sabah* (May 5, 2001) at http://arsiv.sabah.com.tr/2001/05/05/p09.html.

69 Harun Odabasi, "Akkuyu Nukleer Savasi [The Akkuyu Nuclear War]," *Aksiyon*, No. 415 (November 18, 2002) at http://www.aksiyon.com.tr/yazdir.php?id=2131.

absence of a national energy strategy, international concerns over nuclear proliferation, and the domestic political situation.[70]

The coalition government went into a crisis due to the deterioration of Prime Minister Ecevit's health, and decided to hold early elections in November 2002. The election results were much of a surprise for many political observers in that a newly established political party, the Justice and Development Party (AK Parti) formed under the leadership of Recep Tayyip Erdogan, won the majority of the seats in the Parliament with some 35 percent of the votes nationwide, which was more than sufficient for a single-party government.

The AK Parti government reintroduced the issue of nuclear power as one of the major energy sources to reduce supply security risks caused by the dominance of imported fuels, and to ensure diversity in power generation. In 2004, the Energy Ministry revived the nuclear project and launched studies for a long-term and comprehensive nuclear power program. Turkey arranged a cooperative relationship with the United States on civilian uses of nuclear energy, ratifying the agreement in parliament on January 14, 2004.[71] A 2004 report by the Turkish Electricity Transmission Company foresaw that between 2011 and 2015, 4,500 MWe of nuclear capacity would be installed on the basis of the assumption that the energy demand would increase by 7.9 percent between 2005 and 2020.[72] At the end of 2005, the government announced the plans to have at least three nuclear power stations with a capacity of 5,000 MW in operation by 2012.[73] In January 2006, Energy Minister Hilmi Guler declared 2015 as the date of the launch of nuclear power in

[70] Kibaroglu, "Turkey's Quest for Peaceful Nuclear Power, p. 34.

[71] Turkiye Cumhuriyeti ile Amerika Birlesik Devletleri Arasinda Nukleer Enerjinin Bariscil Kullanimina Iliskin Isbirligi Anlasması ve Eki Mutabakat Zaptinin Onaylanmasinin Uygun Bulunduguna Dair Kanun [The Law on the Approval of the Agreement between the Republic of Turkey and the United States on the Peaceful Use of Nuclear Power, and the Memorandum of Understanding], No. 5068 (January 14, 2004) at http://www.tbmm.gov.tr/kanunlar/k5068.html.

[72] *Türkiye Elektrik Enerjisi Uretim Planlama Calismasi 2005–2020* [Turkish Electrical Energy Generation and Planning Study 2005–2020], Turkiye Elektrik Uretim Iletim A.S. Arastirma Planlama ve Koordinasyon Daire Baskanligi [Turkish Electricity Generation and Transmission Company, Research, Planning and Coordination Department] (November 2004). The results of the study are at http://www.teias.gov.tr/apkuretimplani/sonuclar.htm#_Toc84405681.

[73] "Ulusal Nukleer Teknoloji Politikasi [The National Nuclear Technology Policy]," *TAEK* (July 2, 2009) at http://www.taek.gov.tr/tr/belgeler-formlar/func-download/10/chk,744c76c8752f373f4ecf4c470d16e99c/no_html,1/.

Turkey.[74] The Ministry foresaw that it was necessary to make new investments of 57 GWe in high scenario and of 40 GWe in low scenario, until 2020.[75] After the meetings on the relationship between energy and climate change, the Energy Ministry reported to the Environment and Forestry Ministry that the most convenient options for sustainable development were renewable energy sources; the efficient use of energy; clean coal technologies; and nuclear energy.[76]

Since 2003, the AK Parti government stated its determination to make use of nuclear power to meet Turkey's energy needs for the future. Nuclear energy was also in the AK Parti's 2002 elections declaration, where it was stated that the Parti's energy policy was "based on cheap and reliable access to energy, a competitive energy market, decreasing the burden on public budget, and the preservation of the environment and human health. Their rationale was to provide a major alternative or a substitute to the natural gas-operated power plants, which were basically dependent on foreign sources, and with an aim to observe environmental safety."[77]

After assuming power, the AK Parti included this policy in its government program: to avoid energy shortage, "environmentalist nuclear energy sources"[78] would be introduced, along with energy resource diversification.[79] The official statement of the Energy Ministry explained the rationale behind the nuclear energy project on the basis of the criteria that made up Turkey's energy strategy: that nuclear energy was cheap and environmentally friendly, which would boost Turkey's development and help establish the high-technology products and infrastructure.[80]

[74] The Energy Minister Hilmi Guler's speech on the Ministry's 2007 Budget to the TGNA General Board (December 20, 2006), p. 34, at http://www.enerji.gov.tr/yayinlar_raporlar/2007_Butce_Konusmasi.pdf.

[75] "The 2006 Performance Report of the Energy Ministry," Republic of Turkey, Ministry of Energy and Natural Resources (2006).

[76] Ibid.

[77] Adalet ve Kalkinma Partisi Secim Bildirgesi [The Election Statement of the Justice and Development Party], Part 6 (September 26, 2002) at http://www.belgenet.com/secim/bildirge/akp2002–4.html.

[78] See Erdogan's remarks: TGNA General Board Minutes, 22th Term, 1st Legislative Year, 49th Session (March 18, 2003).

[79] The Program of the 59th Government submitted by Prime Minister Recep Tayyip Erdogan to the Turkish Grand National Assembly (March 18, 2003) at http://www.basbakanlik.gov.tr/docs/hukumetprgtr.doc.

[80] "Nuclear Energy," Republic of Turkey, Ministry of Energy and Natural Resources

The Energy Ministry and TAEK prepared a "National Nuclear Technology Policy." In addition to the utilization of domestic resources for energy supply, they foresee the establishment of a 5,000 MWe nuclear production capacity until 2015 with reactor technologies using natural uranium and pressurized water reactors.[81] The policy aimed at the formation and development of the legal, institutional, industrial and personnel infrastructure in accordance with up-to-date technologies. It also included support for research and development activities in the nuclear field, and at the same time would encourage the participation of the private sector in nuclear technology investments.[82]

Regarding the applications of nuclear technology, the policy sought to widen its uses to fields like medicine, industry, agriculture, animal husbandry, environment and food. As the raw material for nuclear fuel, uranium and thorium would be produced out of domestic reserves in order to decrease dependence on foreign sources. To address the issue of radioactive wastes, studies would be carried out to minimize these wastes in order to decrease their impact on human and environmental safety. A "Nuclear Technology Center" would be established to carry out scientific and technical studies, which would include research reactors and pilot facilities. In order to raise the human capital to apply the nuclear technology policy, nuclear training centers would be set up. A "National Nuclear Coordination Board" would be established in order to effectively sustain nuclear and radiation safety in all applications of nuclear technology. In addition, to develop the measures, legal and R&D infrastructure would be ameliorated.[83]

With regard to nuclear infra and superstructure, one should also underscore the significance of the learning process of TAEC/TAEA and TEK/TEAS over the past three decades of the long deliberations for nuclear technology transfer to the country. Turkish scientists, technicians and administrators have accumulated a good deal of knowledge and experience in the nuclear

(May 20, 2009) at http://www.enerji.gov.tr/index.php?dil=tr&sf=webpages&b=nukleerenerji&bn=224&hn=224&nm=384&id=388.

[81] When the policy was announced, the tender process was still in effect. The specification for the use of natural uranium or pressurized water reactors would point at some firms and exclude others. Hence, it contradicts measures of TAEK for bidding.

[82] "Ulusal Nukleer Teknoloji Politikasi [The National Nuclear Technology Policy]," Republic of Turkey, Ministry of Energy and Natural Resources/TAEK (2009).

[83] Ibid.

field, and conducted studies, *inter alia*, on recovery of uranium from various districts of the country. Design studies on uranium extractors, UO_2 powder production, as well as the operation of a pilot plant proved successful, with operations starting in late 1986 with 1.5 ton uranium concentrate imported from Canada. Natural uranium and thorium deposits of Turkey are said to be 8,400 tons and 380,000 tons, respectively, of which the former may constitute a basis for three 650 MWe pressurized heavy water reactors.[84] Thorium reserves may also be used for nuclear energy generation provided the necessary technological steps are taken.

Current projects

Following the proposal of TAEK for some eight locations as the nuclear power plant sites, Prime Minister Erdogan announced in April 2006 that the government had chosen Inceburun in Sinop.[85] The Ministry stated that according to the plans, there would be a need for 4,500 MWe of nuclear energy by 2020, and that one reactor was not sufficient in generating this much power.[86] Immediately after the declaration, the Energy Ministry held a two-day nuclear energy summit at the TAEK headquarters in mid-April 2006. The first included representatives of the private sector, and the second engaged some 150 participants from academia, the Foreign Ministry, the Defense Ministry and Gulhane Military Medical Academy in order to discuss the technology and what the nuclear technology center would encompass.[87] The Energy Minister convened the representatives of some leading Turkish firms[88]

[84] Oner, "Turkiye'nin Nukleer Enerji Santralleri Konusunda Ileriye Donuk Politikalari Ne Olmalidir [What Should Turkey's Future Nuclear Energy Policy Be?]," pp. 18–24.

[85] "Nukleer Santralin Adresi Sinop [The Address of the Nuclear Plant is Sinop]," *NTVMSNBC* (April 13, 2006) at http://www.ntvmsnbc.com/news/368946.asp.

[86] "Ilk nükleer santral Sinop'a kurulacak [The first nuclear power plant to be built in Sinop]," *Milliyet* (April 13, 2006) at http://www.milliyet.com/2006/04/13/ekonomi/eko02.html.

[87] "Nukleer Enerji Zirvesi'nin ikinci ayagi basladi [The Second Part of the Nuclear Energy Summit Starts]," *Sabah* (April 13, 2006) at http://arsiv.sabah.com.tr/2006/04/13/eko134.html; "Guler: Nukleer enerji mecburiyetimizdir [Guler: Nuclear Energy is Our Obligation]," *Hurriyet* (Istanbul: April 14, 2006) at http://hurarsiv.Hürriyet.com.tr/goster/haber.aspx?id=4257133&tarih=2006-04-14.

[88] Koc, Sabanci, Zorlu, Enka, Calık, Gama, Akkok, Ak Enerji, Alarko, Tekfen, Dogus, Guris, Nurol and Habas.

to discuss the plans to build nuclear power plants. In addition to the private sector, energy and economy bureaucracy were also present in the meeting. The talks focused on the investment models and concerns over the possible risks due to the use of nuclear energy.[89]

The AK Parti government also worked on the legal infrastructure for licensing[90] and nuclear fuel deals,[91] and allocated funds for research on domestic technology.[92] The most important issue was financing.[93] The law on the establishment and operation of nuclear power plants passed through the parliament in early May 2007. However, the government and the opposition party were at odds during the deliberations.[94] It was first vetoed by the President of the Republic, Ahmet Necdet Sezer. Law 5710 was passed after incorporating proposed changes, and was issued in the *Official Gazette* on November 21, 2007.[95] The law defined the process of competition (tender) and the selection process for the nuclear power plants, the determination of the establishing firm, and the principles of the sale of the electricity generated. The law foresaw that TETAS would purchase the electricity on the basis of an agreement with the firm. The law also formed two different funds for costs separated in the management of radioactive waste and in the decommissioning of the power plant.[96]

While the licensing work was continuing for the Sinop site, the government opened the tender for Akkuyu, which already had a license, and on September

[89] "Holdings discuss nuclear energy in Ankara," *Turkish Daily News* (April 14, 2006) at http://www.turkishdailynews.com.tr/article.php?enewsid=40767.

[90] "Nukleer enerji icin lisans hazirligi [Preparations for Licensing for Nuclear Power]," *NTVMSNBC* (April 1, 2007) at http://www.ntvmsnbc.com/news/404325.asp.

[91] "Nukleer santralde yetki EUAS'ta [The Power Rests with EUAS in Nuclear Plant]," *Sabah* (July 28, 2006) at http://arsiv.sabah.com.tr/2006/07/28/eko137.html.

[92] Sezer Kilic, "Nukleer için TAEK'e 3 milyar YTL kaynak [3 billion YTL to TAEK for nuclear]," *NTVMSNBC* (March 8, 2007) at http://arsiv.ntvmsnbc.com/news/402183.asp.

[93] "Private sector wants to include public sector in nuclear plants," *Turkish Daily News* (May 12, 2006) at http://www.turkishdailynews.com.tr/article.php?enewsid=43216.

[94] "Nukleer Yasasi sessiz sedasız geçti [The Nuclear Law Passed Quietly]," *NTVMSNBC* (May 9, 2007) at http://www.ntvmsnbc.com/news/407510.asp.

[95] "Nukleer Güc Santrallarinin Kurulmasi ve Isletilmesi ile Enerji Satisina Iliskin Kanun [The Law on the Establishment, Operation and Energy Sales]," *Official Gazette*, No. 26707 (November 21, 2007) at http://www.resmi-gazete.org/tarih/20071121-1.htm.

[96] Soner Aksoy, "Nukleer Enerji Yatirimlarini da Ozel Sektor Yapmali [The Private Sector Should Invest for Nuclear Power]," *İşveren [The Employer]* (December 2007) at http://www.tisk.org.tr/isveren_sayfa.asp?yazi_id=1932&id.

24, 2008 some six consortiums responded, yet only one of them filed a proposal, the Atomstroyexport-Inter Rao-Park Teknik consortium. Those not entering the competition presumably were dissatisfied with guarantees and subsidies, and had expected Treasury guarantees. The Atomstroyexport-Inter Rao-Park Teknik consortium proposed to establish four units of VVER-1,200 (AES-2006) design reactors, which is the Russian type pressurized water reactor. The nominal electrical power of each unit proposed for Akkuyu would be around 1,200 MWe, and the total power of the nuclear power plant composed of four units would be approximately 4,800 MWe.

On December 19, 2008 TAEK announced that the evaluation of the proposal was complete and confirmed that the proposal met the criteria.[97] It was then submitted to TETAS, the authorized institution to open the third letter submitted by the consortium, which contained the price per KWh. On January 19, 2009 TETAS declared that the Turkish-Russian consortium's offer was 21.16 cents per KWh, which far exceeded the expectations for an economical energy investment.[98] Although it was acknowledged that nuclear power plant investments were expensive, this price tag was unaffordable for the government and did not meet the policy criteria of cost-effectiveness. The acceptable price would be in the range of 10–12 cents. In late 2009, the Council of State halted the execution of some articles of the nuclear power plant tender regulation, and the tender for the Akkuyu plant was canceled.[99] Later on, the government decided to sign bilateral agreements at the governmental level for nuclear power plants, that is to say, without issuing tenders. The first one was signed with Russia for a power plant in Akkuyu,[100] and then with South Korea

[97] "TAEK Olcutleri Uygunluk Degerlendirmesi Tamamlandı [The Evaluation of Compatibility with TAEK Criteria is Complete]," *TAEK* (December 18, 2008) at http://www.taek.gov.tr/tr/basin-aciklamalari/41–2008/213-basin-19aralik2008.html.

[98] "Nukleer Santralde Yuce Divanlik Skandal [Scandal in Nuclear Power]," *Hurriyet* (Istanbul: April 24, 2009) at http://www.Hürriyet.com.tr/ekonomi/11505171.asp?gid=229.

[99] "Nukleer Ihalesi Iptale Gidiyor [The Nuclear Tender about to be Cancelled]," *NTVMSNBC* (November 16, 2009) at http://www.ntvmsnbc.com/id/25021766/; "Nukleer Santral Ihalesi Iptal Edildi [The Nuclear Power Plant Tender was Cancelled]," *CNN Türk* (November 20, 2009) at http://www.cnnturk.com/2009/ekonomi/genel/11/20/nukleer.santral.ihalesi.iptal.edildi/552540.0/index.html.

[100] "Russia signs deal to build Turkey's first nuclear power plant," *RiaNovosti* (August 6, 2009) at http://en.rian.ru/business/20090806/155747143.html.

for another power plant in Sinop.[101] The agreement that was signed with Russia was approved by the Turkish parliament on July 15, 2010, and approved later on by the Russian Duma in October 2010 and finally signed by Russian President Medvedev in November 2010.

However, talks with the South Korean government are said to have entered deadlock, mainly due to the lack of funds that would be necessary to sponsor the US$20 billion worth project, as well as other technical issues such as long-term waste treatment and decommissioning of the plant some sixty years after starting operation. South Korea is more than excited to undertake the unprecedented bids for very long-term commitments that Turkey insists are part of the package deal. However, since neither country is currently capable of financing the project, it was agreed that a joint partnership firm would seek potential sponsors from within the international financial markets. Due to the economic crisis at the global scale, no sponsors have been found yet. Turkish Energy Minister Taner Yildiz told the reporters in Seoul during the G-20 Summit in mid-November 2010 that despite the initiatives taken by Turkey's President Gul and Prime Minister Erdogan, it was not possible to find a breakthrough in the deadlock and thus Turkey would consider starting to discuss the bid for the Sinop power plant with other potential suppliers such as Japan.[102] Following this statement of Minister Yildiz, Japanese Ambassador to Ankara Nabuaki Tanaka told reporters that "Japan was ready to provide a nuclear power plant to Turkey from A to Z."[103]

The Undersecretary of the Ministry of Energy, Metin Kilci, was reported to have said on February 1, 2011 that "Turkey aspired to build 20 nuclear reactors by the year 2030, even though that has not been officially stated at any level in the government circles or in the state mechanism and that it was a target and an ideal that Turkey deserved."[104] Kilci has also unequivocally

[101] "Turkey, South Korea sign protocol to establish nuclear power plant," *Anatolian News Agency* (March 10, 2010) at http://www.aa.com.tr/en/turkey-south-korea-sign-protocol-to-establish-nuclear-power-plant.html.

[102] "Nukleer Santralde G. Kore ile Anlasma Yok [No Agreement with S. Korea on the Nuclear Plant]" *EurActiv* at http://www.euractiv.com.tr/enerji/articlenkleer-santracle-/gkore-ile-anlama-yok-013353.

[103] http://www.cnbce.com.

[104] "Enerji Bakanlığı Müsteşarı Metin Kilci, Türkiye'de 2030 yılına kadar en az 20 nükleer enerji reaktörünün devreye girebilmesini hedeflediklerini söyledi [The Undersecretary of Ministry of Energy said they target putting in operation at least 20 nuclear reactors by the year 2030]" at http://istenhaber.com/2011/02/01/turkiyenin-2030-hedefi-20-nukleer-santrale-ulasmak.

stated that "after Japan [was presumably] building the [second set of four] reactors, the Energy Ministry does not envisage to be satisfied with four plus four nuclear power reactors."[105] At a press conference on March 12, Energy Minister Taner Yildiz said that construction in Akkuyu could begin in the next three months. He also stated the country "will accelerate" the estimated seven-year construction period "as much as possible." Days after the Fukushima disaster, Prime Minister Recep Tayyip Erdogan confirmed to the press the country's commitment to build this first reactor, and said the nation would have three functioning nuclear power plants by 2023.[106] According to Energy Ministry officials quoted (but unnamed) by *Reuters* on April 5, 2011, "the third reactor site is considered, though yet to be officially pronounced, to be Igneada in the Thrace region" in northwestern part of Turkey, which is relatively closer to Istanbul.[107]

Responding to questions and criticisms about Turkey's determination to continue with the nuclear projects uninterrupted only days after the Japanese Energy Minister Yildiz told reporters on March 14, 2011 that "the nuclear leakage in the Fukushima Daiichi nuclear reactor will not negatively affect the nuclear plant that will be built in Turkey."[108] Minister Yildiz went on to say that "the Fukushima plant [was] a first generation power plant and the power plant planned to be built in Turkey [would be a] third generation plant."[109] These and other similar statements made by the Turkish Prime Minister as well as the Minister of Energy and his staff indicate the determination of Turkey to exploit nuclear energy at a large scale in the foreseeable future. Turkey seems to be committed to making up for the decades taken in entering the nuclear age. How Turkey's journey will continue from today into the

105 Ibid.

106 Julia Harte, "Building of Turkey's First Nuclear Plant, Sited on a Fault Line, Facing Fresh Questions," *Reuters* (March 25, 2011) at http://www.reuters.com/article/2011/03/25/idUS122778134920110325.

107 "Akkuyu ve Sinop'un ardından üçüncü nükleer santral için Trakya'da İğneada öne çıkıyor" [Igneada in the Thrace comes to the fore as the third nuclear reactor site after Akkuyu and Sinop] at http://istenhaber.com/2011/04/05uncuncu-nukleer-/santral-marmaraya.

108 "Enerji Bakanı Taner Yıldız, Japonya'nın Fukushima Daiichi nükleer santralinde meydana gelen sızıntının, Türkiye'de kurulacak santrali olumsuz etkilemeyeceğini açıkladı" [Nuclear leakage in the Fukushima Daiichi nuclear reactor will not negatively affect the nuclear plant that will be built in Turkey] at http://istenhaber.com/2011/03/14/fukushima-1-nesil-santral-turkiyede-3-nesil-kurulacak.

109 Ibid.

future will also set examples and precedents for actual and potential followers in various parts of the globe.

Conclusion

Nuclear energy is, by its very nature, of dual character. For this reason, a country's bid for nuclear power is usually met with suspicions and much skepticism regarding its true intentions. Therefore, the supplier nations are used to placing a special emphasis on assurances from the recipient countries that there will be no diversion from peaceful to military purposes while using those technological capabilities either acquired through transfers from other states or indigenously developed in later stages. One way to provide assurances to both the supplier states and the international community is to comply with the nuclear non-proliferation arrangements. Yet, in the particular case of Turkey, which has an outstanding record of commitment to the principles and norms of the non-proliferation regimes of all sorts, it is clearly seen that even this much may not be enough. The selective attitude of the supplier nations has much retarded Turkey's aspirations to building nuclear power plants for nearly half a century since the early 1960s. That said, there are certainly lessons that could be drawn from Turkey's experience in its dealings with Western supplier nations.

The first lesson would be to take steps to adhere to all of the nuclear non-proliferation arrangements without delay. In that respect, Turkey's ratification of the NPT, which came some eleven years after its signature in 1969, may have raised the preliminary questions in certain circles that are sensitive about the intentions of states of Turkey's caliber, having powerful military and ambitious politicians. Moreover, Turkey's application for membership in other arrangements designed for export controls could have been made earlier. This would of course require awareness of the significance of being party to these institutional frameworks.

A second lesson is the necessity of a popular consensus within the country—after substantial deliberations in various discussion platforms with the participation of government officials, politicians, academics, community leaders, NGOs and other interested parties—about the various effects as well as the implications of bringing nuclear technology into the country, whereby the undesirability of nuclear weapons ambitions would be agreed upon by the majority of the population and then signed and sealed unequivocally. Sweden's decision to put a binding clause in the constitution would be the ideal situation in order to provide powerful assurances that the country would live up to its non-proliferation commitment for as long as possible.[110] No such comprehensive debate has ever taken place in Turkey except for the routine discussions on the pros and cons of nuclear energy with respect to the environmental effects, the financial scale of the investment, and the like. The proliferation dimension has usually been omitted in Turkey, due both to a lack of awareness about the subject matter in the public domain and to the deliberate attitude of government officials to avoid discussing such "sensitive" issues publicly.

The third lesson would be for the ruling elite of the country, in its political, military and diplomatic circles, to avoid issuing or make irresponsible statements that may create serious concerns about the long-term ambitions of the country. A significant number of former Turkish government officials, as well as retired military personnel, have been on the record arguing that Turkey should consider having nuclear weapons of its own, or at least the capability. These and other such statements have been picked up selectively by foreign intelligence services and then forwarded to their headquarters thereby affecting the stance of their governments in continuing to block Turkey's acquisition of nuclear technology.

The fourth lesson to be noted here is to avoid considering investing in nuclear technology as a policy that would serve as a deterrent against other states that have already developed, or are in the process of developing, nuclear capabilities; or as a policy that would also serve to mobilize popular support to the government in power because of the prestige associated with nuclear technology. Once the message is subconsciously sent to the public that eventually nuclear capabilities might help protect the regime or the government against its rivals, that message might soon develop into substantial expectations for the government to advance in that direction, which would therefore be detrimental to that government or regime when it eventually fails to do so.

Nuclear energy is both too expensive and too dangerous to play political games with. There are arguments against nuclear energy due to its potential hazards for the environment in the case of accidents or due to problems of waste treatments. There are also views endorsing substantial uses of nuclear energy, due to the possibility of large-scale electricity generation at reasonable prices and without carbon emissions. While that debate is not necessarily the primary concern of this chapter, what is emphasized here is that if a country is determined to benefit from peaceful applications of nuclear energy, the

[110] In this respect see Mitchell Reiss, *Without the Bomb: The Politics of Nuclear Nonproliferation* (New York: Columbia University Press, 1988).

decision must be clear and unique. There should be no secondary considerations in the minds of the decision-making elite. Only by adhering to this principle will the world eventually achieve the goal of eradicating the proliferation predicament.

8

Israel and the Invention of Nuclear Opacity

Avner Cohen

On March 6, 1963, the head of the Office of National Estimates at the CIA, Sherman Kent, issued an eight-page memorandum to CIA Director John McCone entitled "Consequences of Israeli Acquisition of Nuclear Capability."[1] The memo viewed the consequences of Israeli nuclearization as grave. "Israel's policy towards its neighbors would become more rather than less tough... it would ... seek to exploit the psychological advantages of its nuclear capability to intimidate the Arabs and to prevent them from making trouble on the frontiers." Furthermore, Kent continued, in dealing with the United States, Israel "would use all is means in its command to persuade the United States to acquiesce in and even to support, its possession of nuclear capability." The Arab reaction would be "profound dismay and frustration," and among principal targets of Arab resentment would be the United States. The Arabs' recourse would be the Soviet Union which would "win friends and influence in the Arab world."

[1] Sherman Kent, "Consequences of Israeli Acquisition of Nuclear Capability," Memorandum for the Director, CIA Office of National Estimates (March 6, 1963). Sourced from NSF, Country File: Israel, box 118, JFKL.

Yet, Sherman Kent had no sense of how, and under what modality, Israel would acquire the bomb. In fact, in an asterisk footnote he notes that "for the purpose of this memorandum, Israeli 'acquisition of nuclear capability' may mean either (a) Israeli detonation of a nuclear device, with or without the possession of nuclear weapons, or (b) the announcement by Israel that it possessed nuclear weapons even though it had not detonated a nuclear device." Israel did acquire nuclear weapons capability in the late 1960s, but it did so in a very different way from what Kent had assumed.

In contrast with the way Kent had conceived the emergence of the Israeli bomb—open, declared, public—Israel developed, acquired and ultimately possessed nuclear weapons through a very different model that Kent conceptualized, the model of *nuclear opacity*. Israel followed neither of Kent's two modalities. In fact, Israel was able to maintain opaque regional nuclear monopoly by deliberately defying the conventional modes of joining the nuclear club—that is, without testing or even acknowledging its nuclear status.

Opacity, ambiguity, vagueness and secrecy were intrinsic to the making of Israel's nuclear posture. This pattern, of course, was not utterly unique to Israel, but in no other country's case did it ultimately become the nation's trademark. Israel is the only nuclear weapons state, among all of today's nine nuclear weapons states which is firmly committed to a policy and posture (not shared even by North Korea) of opacity—total non-acknowledgment and secrecy.

More than half a century after the initiation of the Israeli nuclear program, and more than four decades after Israel crossed the nuclear threshold, it is worthwhile to revisit Kent's forecast. Specifically, there are two sets of questions that are worth exploring. First, how did Israel invent nuclear opacity? Second, what was the regional impact of Israel reaching nuclear monopoly? That is, what has been the impact of Israel's regional nuclear monopoly on the dynamics of the Arab-Israeli conflict, in particular on the dynamics of war and peace in the region over nearly a half a century?

This chapter is divided into two basic parts. First, it narrates historically the formation of Israel's unique posture of opacity, to be contrasted against the "standard" model by which other states acquired nuclear weapons. Second, it carefully examines the impact of Israel's opaque nuclear monopoly on the dynamics of the Arab-Israeli conflict, and in particular on issues of war and peace in a regional context.

More than half a century after the initiation of Israel's nuclear program, and more than forty years after Israel crossed the nuclear threshold, our historical understanding of the impact of Israel's nuclear weapons program on the region is still murky, if not fundamentally lacking.[2]

The formation of Israel's nuclear opacity

Israel, as noted, is the only nuclear weapons state committed to a policy of opacity. Even if all the primary historical sources were available to us, it would still be difficult to construe a fair—let alone complete—account of those decisions, due to the veiled and opaque decision-making style. But, of course, almost all the primary sources for Israel's nuclear history—documents and human experience—are still unavailable; they are lacking, silent, or at best incomplete.

Still, there are reasons to think that a few key junctions had an extraordinary impact on the entire evolution of Israel's nuclear posture and style. Those early decisions—each could be viewed as a response to a certain nuclear dilemma—generated legacies that shaped the way Israel acts and thinks on nuclear matters to this day. The story of those decisions is the tale of how Israel created itself as a different kind of proliferator. These were critical points in the making of Israel's special bargain with the bomb.

From the start Israel's nuclear development was ambivalent in nature. As much as Israel's nuclear pursuit was characterized by resolve in the technological arena, it was typified by political caution. Israel's technological commitment was vigorous, but its politics was hesitant, tentative and non-explicit. The pattern by which Israel made its most fundamental decisions in the nuclear field highlights this schizophrenic calibration between these two opposing forces, technological resolve and political caution.

Project initiation

The first dilemma that every potential nuclear proliferator must consider is project initiation: deciding the desirability and feasibility of the nuclear program. The decision to initiate a nuclear program is the first decision on the

[2] See Avner Cohen, "On the Historiography of the Israeli Nuclear Project" (In Hebrew), *Israel*, Vol. 9 (2006): pp.245–54. The author spent nearly a decade researching this history but the result, the book *Israel and the Bomb* (New York: Columbia University Press, 1998), captured mostly the external layers of the story—primarily its political and diplomatic aspects—leaving the tale of those decisions almost untouched.

proliferation ladder, but it is a highly important one. Deliberating the initiation issue means inevitably deliberating the entire project. A response to this dilemma tells a great deal about the decision makers' intentions.

When Prime Minister David Ben Gurion made his initiation decision to kick off a national nuclear project he unequivocally chose the path of technological resolve. The technological package that Israel generated in France in 1956–7 was extraordinary in its scope and was tantamount to a full-blown national commitment to build a weapons infrastructure.[3] Enough is known today about the breadth of the initiation stage to realize how committed and dedicated the early vision was.[4] In essence, the Dimona complex was designed to include all the technological components required for a plutonium-based nuclear weapons infrastructure.

The scope and determination of the vision were manifested in the *sanctum sanctorum* of the facility—the deeply underground reprocessing facility designed to extract plutonium. Nothing is more indicative of Israel's commit-

[3] The best account of the French side of the Dimona deal is still Pierre Péan, *Les deux bombes* (Paris: Fayard, 1981). In one memorandum of understanding that Peres signed with French Defense Minister Maurice Bourgès-Maunoury on August 23, 1957, the parties committed themselves to cooperate in the research and production of nuclear weapons. This was probably an improper—if not illegal—document from its birth as both men were not authorized to commit their countries to such objectives, but it surely tells us something about Israel's nuclear resolve. Michael Bar-Zohar, *Shimon Peres: The Biography* (New York: Random House, 2007), pp. 211–12. This memorandum of understanding was most likely signed without the knowledge of the French cabinet, which at that point had not yet made any formal decision on the bomb. See also *International Herald Tribune*, "Author says Shimon Peres persuaded France to backdate nuclear deal with Israel in 1957," March 20, 2007. That memorandum, which was frozen shortly after it was signed and subsequently annulled, but no one could be mistaken about the Israeli long-term intentions that it reflects.

[4] From the outset the initiation vision involved setting up the organizational structure for the two aspects required for a nuclear weapons complex: the plutonium-based production facility at Dimona and a central weapons lab to be part of a newly organized defense R&D authority (RAFAEL) to be located in the north of Israel. Both organizations were set up in 1958, with both reporting to the project chief, Shimon Peres. The driving idea behind this organizational structure of a "house divided" was security and control. Each side in the divided project should focus on its job with as little direct contact as possible, until such contact became imminent and necessary. Over time this organizational structure would become increasingly inefficient and costly, to the point of becoming almost a threat to the mission itself, but initially this was a way to ensure security even in case of premature disclosure.

ment to technological resolve than this super secret and costly facility. It was hoped that within a decade or so Israel would be in a position to build its own nuclear device.

Ben Gurion was aware of the risks involved in his initiation decision, but they did not deter him.[5] Instead those risks shaped the *modus operandi* under which Israel pursued the project. The initiation decision was essentially a long-term proliferation decision, but Ben Gurion apparently presented it—even to his close political colleagues—as if he was only hedging against the potential failure of other short-term strategies of defense.[6] Presenting this distance between the initiation issue and its long-term strategic implications served a purpose. For, even though such a distance was artificial—after all, decisions about nuclear infrastructure make little sense without some reference to long-term objectives—it was beneficial politically to avoid making long-term commitments.[7] Viewed in this way, the initiation of the Dimona project was presented as if it was about creating "options" for "future leaders," not about a commitment to build the bomb.[8]

Years later, in a moment of historical reflection, Shimon Peres referred to Ben Gurion's *modus operandi* as reluctance to "nail down" the specifics of his nuclear vision, "for nailing down would have meant to identify specific objectives too early, and too fast, and that would have been too complicated."[9] Those long-term objectives were left vague and undecided. Evidently, the commitment to technological resolve was moderated (at times even disguised) by political caution. This *modus operandi* was critical in Israel's path to opacity.

[5] Avner Cohen, *Israel and the Bomb* (New York: Columbia University Press, 1998), pp. 63–7. During the initiation stage Israel became highly dependent on the good will of the foreign supplier, France, to carry out its commitment. Had France pulled out at an earlier stage without reaching a compromise, the entire project could have collapsed. An early discovery of the project by the United States could have also killed the project.

[6] Ariel Levite, "Never Say Never Again: Nuclear Reversal Revisited," *International Security*, Vol. 27, No. 3 (2002–3): pp. 59–88.

[7] As a matter of speculation, it is doubtful if Ben Gurion shared with even his closest ministerial colleagues information about the underground reprocessing plant under construction—the core of the project.

[8] Cohen, *Israel and the Bomb*, pp. 76–8.

[9] Shimon Peres, "About Shalheveth" in *Shalheveth Freier, 1920–1994*, Israel Atomic Energy Commission (IAEC) (Tel Aviv: 1995).

The disclosure dilemma

The second nuclear decision that Israel had to face was on a strategy of how to present its nuclear project to the world. Throughout 1960 it became increasingly evident that the secret nuclear project could no longer remain a total secret, and Israel would have to address it publicly. There were two available choices for dealing with the dilemma: Israel could adopt a cautious strategy, claiming that the project is strictly civilian and peaceful; or, alternatively, it could endorse a more bold approach, acknowledging that the project was about defense. Whatever was said, however, it did matter. It had a profound impact on what Israel could or could not do (or say) in the future.

But the disclosure dilemma had a context. Already in early 1960, before US intelligence knew much about Dimona, Israel was in a crisis with France over the nuclear issue. After France changed national leadership it turned sour on its "unprecedented" nuclear deal with Israel. France's new President, Charles De Gaulle, decided to end the nuclear partnership with Israel and to come clean. Not only did he decide to halt French involvement, but he insisted that Israel must declare that the Dimona project was peaceful and place it under safeguards.[10] A hastily arranged summit between Ben Gurion and de Gaulle produced no solution.[11] In Israeli eyes, this was a fight over the survival of the Dimona project. At stake was Israel's commitment to nuclear resolve.

From the perspective of Shimon Peres (Deputy Minister of Defense), the impasse with France was not a policy dilemma but a challenge of necessity. Peres sought to devise a compromise allowing Israel to complete Dimona on its own without French governmental involvement. Two issues were critical. First, certain kinds of French technical assistance were indispensable. Second, secrecy on some aspects of the project needed to be maintained. The kind of compromise Israel could achieve with France was also critical for what Israel could say later to the United States. If no compromise could be found, and if the French were to "tell it all," this could be catastrophic for Israel's relationship with the United States.[12] In late November, Israel found a solution. Shimon Peres told the French that Ben Gurion would soon issue a public statement about Dimona, pledging that the project was peaceful.[13] According

[10] Michael Bar–Zohar, *Shimon Peres: The Biography* (New York: Random House, 2007), p. 228.

[11] Ibid., p. 231.

[12] Ibid., pp. 231–4.

[13] Interestingly, this detail appears only in the Hebrew edition of Bar-Zohar's biogra-

to Peres' biographer, this compromise "saved the Israeli nuclear project from a total disaster."[14]

A few days later Dimona was revealed. The high-profile American disclosure, along with the press' speculations that Dimona meant an Israeli bomb project, created a real dilemma for Israel—how should Israel respond to the disclosure? How much should Israel tell? Could Israel tell the US government in private different things from what it told the world in public? Should Israel make peaceful commitment for the future? Simply put, should Israel choose the bold path of truth or the cautious path of covering up the truth?

Ben Gurion took the path of caution. On December 23, 1960, he delivered the first (and only) public statement on this matter in the Knesset (the Israeli legislature), in which he referred to the Dimona project as a strictly civilian nuclear research center. The Dimona reactor, Ben Gurion stated, was "exclusively for peaceful purposes." In addition, the Israeli government privately provided the United States with additional information about the nuclear project that was consistent with the public statement.[15] The United States took it as an Israeli commitment that "Israel has no plans for the production of atomic weapons."[16] Neither the public statement nor the additional clarifications made even a veiled reference to the security-related dimensions of Dimona.

Ben Gurion's public statement was not the whole truth, to say the least, but at stake was not truth but politics.[17] Ben Gurion apparently decided to adopt the cautious strategy for its short-term benefits: avoiding a public showdown with the US, lessening American pressure on Israel, and creating a political lull. A showdown with the United States over Dimona would, it was feared, be catastrophic.[18]

---

phy. Michael Bar-Zohar, *Shimon Peres: A Political Biography* (In Hebrew) (Tel Aviv: Yediot Achronot, 2006), p. 326.

[14] Ibid., p. 234.

[15] On practical issues, however, Ben Gurion was more cautious. He evaded the question of the ownership of the plutonium and rejected formal international inspection of Dimona. He accepted a visit by American scientists, but made it clear that this would be carried out under Israeli control.

[16] Cohen. *Israel and the Bomb*, p. 95. See also Jeremy Salt, "Israel's Nuclear Weapons: The White House Factor," *Middle East Policy*, Vol. 17, No. 3 (Fall 2010): pp. 24–5.

[17] The real rationale and purpose of the Dimona complex has nothing to do with peaceful scientific research. It has all to do with the purpose of producing plutonium.

[18] Apparently for this reason, Ben Gurion postponed his resignation for two months.

But those short-term benefits need to be weighed against their long-term fallout. For one thing, declaring that the Dimona project was "exclusively peaceful" while secretly continuing to work on its true mission set Israel on a future route of deception with the United States. Such a strategy opened the door for more American demands to verify the "exclusively peaceful" pledge, such as additional requests to visit the Dimona site and, ultimately, demands to place Dimona under international safeguards. If Dimona was "exclusively peaceful," why worry about visitors or safeguards?

Moreover, the logic of this cautious strategy made it difficult for Israel to reap the security benefits from its nuclear investment. Even if Israel was prudent not to be caught in its untruth, it was likely to end up with a secret nuclear capability that could not be utilized for its security benefits. Since this disclosure strategy did not make it possible to present Dimona as an asset to Israel's security, Israel would have difficulties in the future using Dimona's potential to bargain for some other security assets. Most prominently, this strategy rendered it nearly impossible for Israel to join the nuclear club at a later time.

In retrospect, Ben Gurion's endorsement of this disclosure strategy had a lasting effect. It placed major political and technological inhibitions on Israel's freedom of action in the nuclear field. For, after Ben Gurion's address in the Knesset, Israel was formally committed to the "peaceful use" of its nuclear technology. Moreover, Israel was forced to grant visits to verify its peaceful commitment and was left to pursue its nuclear program under a path of concealment and deception. Israel was not even in a position to demand American security guarantees for concession on the nuclear issue.

Ben Gurion's response to the disclosure dilemma in December 1960 was eclipsed by subsequent decisions, but this initial declaratory strategy shaped the entire political space in which those subsequent decisions were made. This strategy determined a great deal of the parameters that governed US-Israel nuclear relations throughout the 1960s and beyond. Under the shadow of this strategy much of the American-Israeli nuclear dialogue throughout the 1960s took place: the Kennedy-Ben Gurion meeting in May 1961, Kennedy's letters to Ben Gurion and Eshkol in the spring and summer of 1963, the Kennedy-Eshkol deal in August 1963 that led to annual US visits to Dimona, and

---

In December 1960 he had been in the midst of a domestic crisis over the Lavon affair, and he was determined not to resign from the government under these circumstances.

Eshkol's 1965 "non-introduction" pledge. Ultimately, the "non-introduction" commitment was at the heart of the Nixon-Meir deal of 1969.

While American efforts to halt Israel's nuclear resolve ultimately failed, they helped to shape the cautious and opaque mode under which Israel became a benign nuclear weapons state. The United States was not in a position to reverse Israel's nuclear resolve, but it was powerful in determining the parameters of caution under which Israel went nuclear. In addition to the role of the United States, Israel's disclosure strategy was an equally important junction on its way towards opacity. From early on the impulse for technological resolve was checked by a nearly equal impulse for political caution.

Some may argue that it is unfair to treat the disclosure decision as a real dilemma because Ben Gurion had real alternatives to the course of action he chose. Even Ben Gurion's own Foreign Minister, Golda Meir, saw room for alternatives. She held the opinion that Israel should "tell [the United States] the truth, and explain why." Yet, even if this was the case—and we do not really know how Ben Gurion saw things—had Israel adopted that policy, it is likely that Israel would have ended up in a very different nuclear posture.

Nuclear versus conventional

The next rung on the proliferation ladder involved the strategic role of the nuclear program within Israel's overall national security posture. In general, there is no exact timeline charting when such deliberations should be initiated, and it varies from one case to another, but logic suggests that such deliberations should take place sometime before the research and development phase reaches completion.

Such high-level deliberations must confront the strategic fundamentals of the nuclear program: the role of the nuclear program within the nation's overall national security posture, the strategic vision that drives the nuclear program, the ratio between conventional and nuclear components, and the specific products or capabilities that the program should produce. Basic issues such as national security doctrine, deterrence, force structure, military utility of nuclear weapons, defense budgets, civil-military relations, organizational structure, and the like must all be introduced into the overall picture. Those deliberations are never final—there is an essential sense of tentativeness about them—but once a national consensus is formed it provides, at least in theory, lenses through which to examine the program.

Israel was decisive on the issue of the infrastructure, but hesitant, slow, and tentative on the next sets of decisions. This reflects Ben Gurion's basic pattern in nuclear affairs: his reluctance to make long-term decisions, his tendency to leave nuclear intentions ambiguous, undefined, and unstated. There were also concrete reasons to his delay on long-term issues. For one thing, in 1960–1 the survival of the Dimona project was in limbo due to the reverse in policy by France and the Kennedy administration's opposition to the project. For another thing, the cautious declaratory strategy that Ben Gurion adopted made it difficult to even bring the topic up for negotiation. On the domestic side, Ben Gurion's leadership was bitterly challenged in generational infighting within his own party (known as the "Lavon affair") that consumed his political energy and eventually led to his downfall. Given all this, from Ben Gurion's perspective, this was surely not the time to start raising long-term sensitive questions about the nuclear project.

But new developments in 1962, abroad and at home, forced the issue on him. On July 21, 1962, Egypt announced the successful test firing of four ballistic missiles, proclaiming the existence of a missile project aimed to hit any target "south of Beirut." Israel was stunned. The Mossad issued an alarming report on hundreds of German scientists and engineers who were developing in Egypt a variety of technological projects, including ballistic missiles and unconventional weaponry. The Israeli media were spreading horror stories whose central theme was that fifteen years after the Holocaust German scientists were working on weapons to bring another Holocaust to the Jews.[19] Almost overnight Israel was engulfed in a debate over what its strategic response to the news from Egypt should be.[20]

The leaders of Achdut Ha'avodah—the junior coalition party—Yigal Allon and Yisrael Galili, had their own concerns about the atomic project. They asked to discuss the fundamentals of the nuclear issue with Ben Gurion and to register their concern. They had already, upon joining his government in October 1961, inserted in their coalition agreement a clause that a ministerial

[19] Bar-Zohar, *Shimon Peres: The Biography*, pp. 256–8.

[20] Shimon Peres had serious reservations about the specifics of the Mossad's alarmist assessments, but he viewed the Egyptian missile project as a new trend towards "deterrent weapons" that he considered historically inevitable. During the second half of 1962 Peres was a vocal advocate for the "new deterrence" school. Without explicitly advocating the bomb—the bomb issue was already a taboo—he argued that Israel must develop new and powerful "deterrent weapons," not to win the war but to persuade the Arabs not to arrive to the "wrong conclusions." Those discussions were so intense that they made the US embassy in Tel Aviv issue a seven-page report on the strategic debate in Israel.

committee must be responsible for decisions on "the development of new weapons' system to be deployed by the IDF [Israel Defense Forces]."[21]

The result was the first closed-door leadership conference in which the strategic fundamentals of the nuclear project were debated by the nation's top politicians with security/defense credentials.[22] To this day the minutes of that meeting, if they exist at all, are unavailable. It is not even exactly clear who attended the meeting. According to Arnan Azaryahu, Galili's aide who was involved in the preparations for that conference, the conference was essentially a debate between two schools of thought, the advocates and the opponents of nuclear deterrence.[23] The meeting took place apparently in the second half of 1962, possibly at Dimona, and both sides of the debate prepared position papers.[24] For contemporary academic audiences, the debate anticipated many of the classical arguments that international security theorists such as Kenneth Waltz and Scott Sagan would debate many decades later on the spread of nuclear weapons.[25]

The advocates of nuclear deterrence were former Chief of Staff and then Minister of Agriculture Moshe Dayan and Deputy Minister of Defense Shimon Peres (who had direct responsibility over the nuclear program). Their

[21] Cohen, *Israel and the Bomb*, p. 148.

[22] There were MAPAI (Workers' Party of the Land of Israel) senior ministers who knew of and even discussed the nuclear project—Treasury Minister Levi Eshkol, Foreign Minister Golda Meir, Trade Minister Pinchas Sapir, and others—but these politicians lacked security/defense credentials and they did not discuss the nuclear issue on largely strategic grounds.

[23] I discussed this meeting in a long interview with Arnan Azaryahu that took place at his home in Kibbutz Yiron on August 22, 1992. I took extensive notes out of that taped interview. I hardly used this interview when I wrote *Israel and the Bomb*, and this is the only occasion that I returned to the contents of that interview.

[24] Interviews with Arnan Azaryahu conducted by the author (Summer 1992 and January 2008). According to another account, the meeting took place under the auspices of the Ministerial Committee on Defense which Ben Gurion chaired. Notably, in 1962 Galili and Allon were the leaders of Achdut Ha'avodah, but they were not ministers in Ben Gurion's cabinet. It is possible that there was more than one meeting. See Moshe A. Gilboa, *Six Years, Six Days: Origins and History of the Six Day War* (in Hebrew) (Tel Aviv: Am Oved, 1968), pp. 29–30; Yigal Allon, *Contriving Warfare* (in Hebrew) (Tel Aviv: Ha'kibbutz Ha'meuchad, 1990), pp. 200, 205, 207, 305; Yair Evron, *Israel's Nuclear Dilemma* (Ithaca: Cornell University Press, 1994), pp. 6–7.

[25] Scott D. Sagan and Kenneth N. Waltz, *The Spread of Nuclear Weapons: A Debate Renewed* (New York: Norton, 1995).

arguments echoed Ben Gurion's own pessimistic outlook on the future of the Arab-Israeli conflict (Peres and Dayan were surely influenced by the thought of the French strategist Pierre Gallois). They started with the presumption that Israel must establish a long-term stable deterrence capability but argued that Israel could not afford to keep up in a conventional arms race over the long run for financial and other reasons. The continuation of the conventional arms race would drain the Israeli economy and prolong the conflict and, Dayan warned, Israel would be unable to maintain stable deterrence for the long run. Their conclusion was that only advanced technological weaponry—atomic weapons and missiles—could provide Israel the deterrence edge it needed for the long run. Not only was the bomb the most effective and least costly deterrent weapon, it had the potential to eventually convince the Arabs to accept the reality of Israel. In the absence of external security guarantees, then, the bomb must be Israel's own ultimate deterrent, they argued; the introduction of the bomb could ultimately bring the Arab-Israeli conflict to its end. They urged that Israel should make bold changes in its overall defense policy—that is, in its military doctrine, budget allocations, force structure, organizational structure, etc.—to move toward an advanced technology army.[26] Evidently, they viewed nuclear proliferation as an inevitable trend of the future in global affairs, and thought that Israel must take the lead. They promoted the notion that the Israel Defense Forces (IDF) should be reorganized "as an army of deterrence and decisive victory by relying on the anticipated achievements in science and technology of the 1970s."[27] They called their advocacy "the doctrine of self-reliance," an Israeli version of the French contemporary notion of *force de frappe*.

The counter view, represented by Galili and Allon, started with an attitude of deep skepticism toward nuclear deterrence based on the superpowers' "balance of terror." They rejected all three fundamental presumptions of the nuclear advocates: questioning the inevitability of the global spread of nuclear weapons, dismissing the view that the bomb was the only cost-effective solution for Israel's long-term security, and, most importantly, raising doubts about the applicability of nuclear deterrence calculus to the context of the Middle East. Although Galili and Allon doubted, contrary to Dayan and Peres, that nuclear weapons would one day spread to arsenals all over the world, they did think that the introduction of nuclear weapons to the Middle

[26] Ibid.

[27] Gilboa, *Six Years, Six Days*, p. 30.

East would lead to inevitable regional proliferation. They maintained that an Israeli nuclear monopoly would be merely short-term, and would be replaced by a situation in which both sides of the Arab-Israeli conflict would have atomic weapons. In short, the conflict would be nuclearized. Even if Egypt were not able to compete with Israel in the nuclear field on its own, the Soviets would not allow Israel to maintain a nuclear monopoly in the region. Given the geopolitical and demographic asymmetries of the Arab-Israeli conflict, Israel's national interest was naturally against the nuclearization of the conflict. A non-nuclear Middle East was strategically much preferable for Israel. Furthermore, investment in a costly nuclear weapons program would weaken the IDF and was likely to encourage the Arabs to wage a preventive war against Israel.[28] Significantly, they did not argue that Israel should not engage in some nuclear and missile research and development (R&D)—they did maintain that Israel should maintain its lead in nuclear R&D over its Arab adversaries—but said Israel needed to be very careful that its own actions would not stir a nuclear arms race in the region. Israel's ultimate interest was to keep the Middle East free of nuclear weapons.

In the absence of public record it is unclear how Ben Gurion concluded the conference, and what, if any, operational decisions were issued. But in the annals of Achdut Ha'avodah the meeting was remembered as their victory in the nuclear debate. Galili and Allon believed that they had persuaded Ben Gurion that a nuclearized Middle East would be bad for Israel, and that the IDF must remain conventional.[29] They believed that they were able to convince Ben Gurion to reject the call to move the IDF toward a nuclear-capable force in the way France was moving its own military toward nuclear capability in those days. From their point of view, they had installed a firewall that would prevent Shimon Peres from realizing his dream to make Israel a member of the nuclear club.[30]

Ben Gurion may have seen things a little differently, more pragmatically and less ideologically. Surely he was not interested in rocking the boat. From his perspective, the purpose of the debate was probably more political than

[28] Yigal Allon, *Curtain of Sand* (in Hebrew) (Tel Aviv: Ha'kibbutz Ha'meuchad, 1968): pp. 400–402; also Allon, *Contriving Warfare*, pp. 195–209.

[29] Interviews with Arnan Azaryahu conducted by the author (Summer 1992 and January 2008).

[30] Interviews with Arnan Azaryahu (Sini) conducted by the author, Kibbutz Yiron (Summer 1991 and January 2008). See also Yair Evron, *Israel's Nuclear Dilemma* (in Hebrew) (Tel Aviv: Ha'kibbutz Ha'meuchad, 1987), pp. 17–18.

operational in the first place. It was a way to let off some political "steam" on a sensitive subject that until then had never been debated strategically by politicians. Ben Gurion probably wanted to assure Galili and Allon that he was not rushing the IDF into a nuclear posture in order to assuage their apprehensions about the project. Because Ben Gurion had initiated the project on his own and without democratic consent, it was probably important to him to maintain a sense of Zionist consensus in the project.

Furthermore, while Ben Gurion was definitely interested in "the bomb" as a national insurance policy, it is plausible to suggest that by 1962 he had already concluded Israel was not in a political position to join the nuclear club. After all, he had given his word to President Kennedy a year earlier that Israel would not acquire nuclear weapons. Besides, Ben Gurion probably still maintained that making fateful and long-term decisions on this subject was premature, theoretical and unnecessary. The nuclear program was too young and uncertain to warrant making any structural changes to the IDF. My own suspicion is that in 1962 Ben Gurion had no desire whatsoever to impose changes on the IDF against the will of its top generals. To preserve the IDF conventional doctrine—of which Ben Gurion himself was the prime author—was the practical and obvious thing to do. These considerations, not strategic considerations against nuclear deterrence, probably led him to support Galili and Allon on this.[31] It was also a way to assure the IDF leadership—which had little interest in nuclear affairs—that the secret project at the Ministry of Defense was not a threat to their own turf.

Interestingly enough, during his last year in office (1962–3) Ben Gurion supported proposals that came from both sides of the nuclear debate. On the side of resolve, Ben Gurion authorized the initiation of an Israeli missile project in cooperation with the French firm Marcel Dassault.[32] Israel had to

[31] Allon, *Contriving Warfare*, p. 305; Moshe A. Gilboa, *Six Years, Six Days*, pp. 29–30.

[32] To this day Israel has not officially acknowledged the history of its ballistic missile program, but references to the missile contract appear in US documents from as early as in mid-1963. In 1996 Marcel Dassault published its own history of the MD-620 project. According to the Dassault official history, Israel commissioned its request for the development of the tactical missile on September 7, 1962 from Dassault Aviation. The original Israeli specification was for a surface-to-surface ballistic missile system capable of delivering a 750kg warhead with a range of 235–500km, with a CEP (circular error probability) of less than 1km. Israel's initial request also specifies a missile system that could launch between four and eight missiles per hour in all weather from either a fixed or a mobile launcher. The request specifies a prepa-

respond to the challenge of an Egyptian missile project with a missile project of its own. This decision inevitably strengthened the nuclear commitment. Israel began developing a ballistic missile that could be tipped with a nuclear warhead. Indeed, despite Israel's non-acknowledgment policy surrounding its nuclear program, the missile project made little sense in a non-nuclear context. While the missile project was not a platform that could change the organizational structure of the IDF, a small missile unit was established at the headquarters of the Israeli Air Force (IAF). Over time that small Air Force unit would liaise with the nuclear project. Decades later it evolved to be a wing base, the home of Israel's missile squadrons.

On the side of caution, Ben Gurion rejected proposals to put more funds into the nuclear project at the expense of the conventional army. Instead, as a show of his continued support to Israel's conventional forces, he authorized creating a new regular armored brigade in the IDF.[33] All doctrinal and organizational issues were postponed, pending the completion of the missile program in the late 1960s.

In a long-term historical outlook one could argue that the 1962 nuclear conference may have been very instrumental in creating important legacies and precedents, both in substance and in form, that shaped the way Israel acted and thought on nuclear matters. One could argue that in retrospect it was a fateful junction in Israel's nuclear history. Here are some of those legacies:

- The IDF would remain a conventional army in its basic orientation (that is, in its force structure and doctrine) as long as Israel confronts conventional threats.
- The nuclear program would be treated as a unique national asset, a sort of national security insurance policy under the control of the prime minister, not as just another military weapons system in competition with other military weapons systems.

---

ration time of two hours, and a success rate of 90 percent. The Jericho-1 design is revolutionary in that it is the first French missile to use an on-board computer. The Jericho-1 (originally designated MD-620) was developed in secrecy by Dassault in cooperation with the French Ministerial Armaments Committee (DMA), but without financial support from the French government. Pierre Langereux, "Dassault lève le voile sur le missile Jericho" [Dassault lifts the lid on the Jericho missile story], *Air & Cosmos/Aviation International*, No. 1590 (December 6, 1996): p. 36. See also: http://www.aeronautics.ru/archive/wmd/ballistic/ballistic/jericho_1–01.htm.

[33] Allon, *Contriving Warfare*, p. 305.

- Israel would be politically committed not to introduce nuclear weapons to the Middle East.
- The utility of the nuclear program was essentially political and psychological, not military.
- The nuclear program would have to be handled with extreme caution by its political and professional custodians and it should not become a vehicle through which to change the IDF, and surely not a platform on which to economize national security.
- Because the nuclear issue needed to be handled with extraordinary secrecy and discretion by the prime minister, and because democratic procedures had to be bypassed, it was critical to maintain national consensus.
- Decisions on the nuclear issue needed to reflect that national consensus in a manner that would not stir a public debate.

It took a while before these insights were incorporated and solidified into a national outlook in a near canonic form—this happened in the mid-to-late 1970s under the inspiration of the Director of the Israeli Atomic Energy Commission (IAEC) General Shalheveth Freier—but I suggest that the 1962 conference was a milestone in stimulating these insights. Even though Ben Gurion apparently left no written legacy on the nuclear issue, it would be historically fair, I think, to treat these insights as part of Ben Gurion's unwritten nuclear legacy. A great deal of this legacy is still alive today.

Substantively speaking, as it turned out, the notion that Israel should shift its IDF conventional doctrine to an overt nuclear doctrine was never seriously considered any more. It became obvious that this would not happen, politically and strategically. Even after the trauma of the 1973 Yom Kippur War, when some prominent Israeli defense intellectuals advocated a certain—but still very limited—role for nuclear weapons in Israel's military doctrine, this advocacy soon faded away.[34] It became an Israeli article of faith that as long as the conflict remained conventional, Israel's military doctrine must remain conventional as well. Israel had no interest in viewing nuclear weapons as mere military weapons. Instead, they were solely at the service of the prime minister who retained them for the possibility of an existential moment of last resort. With respect to process, the 1962 nuclear debate created some important legacies in the way Israel handled nuclear matters. It was the last time that

[34] Moshe Dayan kept a veiled variant of this view alive in some of his writings and speeches before the 1977 elections, but subsequently he abandoned his open nuclear advocacy.

matters involving nuclear strategy were debated in a top-level political forum. From that point on such sensitive issues were discussed primarily by professional forums of experts, not by politicians, and when politicians authorized certain decisions they did so in small sub-committees. Even when Prime Minister Golda Meir made her own fateful nuclear decisions in 1969–70 (discussed below), with direct involvement of virtually the same leaders who had been present in the 1962 conference (for example Dayan, Allon, Galili and possibly Meir) it was not like the debate of 1962. While the nuclear divide between the two camps lasted for many more years—in fact Dayan and Allon saw nuclear matters very differently until the end of their political careers—it produced no more showdowns like the one in 1962.

The post-R&D dilemma and the 1967 War

By 1966, under Prime Minister Levi Eshkol, a new set of outstanding nuclear decisions started to accumulate. They were all related to one key development: Israel was approaching the nuclear weapons threshold as the R&D phase neared completion.[35] The two major partners of the divided project were close to convergence, but that convergence proved to be a thorny process, involving bureaucratic, technological and personnel mismatches.[36] It was imperative to set up a new administration to be in charge of all aspects of the project. In mid-1966, the IAEC was reorganized—after a long period of dormancy—as the new executive body under the leadership of Professor Israel Dostrovsky of the Weizmann Institute as its Director General. Prime Minister Eshkol appointed himself IAEC chairman, making clear that nuclear affairs were the sole responsibility of the prime minister.[37]

These changes were vital to complete the R&D phase, but they could not, in themselves, provide the necessary political guidance. The completion of the R&D phase brought the project to a crossroads. New decisions needed to be made about the depth of Israel's nuclear commitment. But Eshkol was hesitant as to the direction in which he should take the project.[38]

[35] According to French sources, plutonium was being separated at Dimona by 1966, theoretical and experimental design work of the first Israeli nuclear device was at an advanced stage, and the French-Israeli missile under contract with Marcel Dassault in France (designated MD-620) was in its first testing stage.

[36] Interviews with Prof. Yuval Ne'eman conducted by author (1996 and March 2006).

[37] Cohen, *Israel and the Bomb*, pp. 223–4.

[38] The French-Israeli missile project was leaked to the public by US intelligence ser-

For all previous nuclear proliferators, the R&D phase ended with a full-yield nuclear test. Such a test not only demonstrated technical capability, but also indicated that the state had made a nuclear commitment. Testing meant a claim to membership of the nuclear club, a way to acknowledge the state's new international status. As was so evident in the cases of India and Pakistan, the latest additions to the nuclear club, the test meant an opportunity for the government to issue a public statement that clarified the nation's nuclear commitment.

Israel was apparently in a position to conduct a full-yield nuclear test in the second half of 1966, had its leaders so chosen. Had Israel conducted a test that year, even a so-called "peaceful nuclear explosion," it could have declared itself the world's sixth nuclear state, and subsequently it could have joined the Non-Proliferation Treaty (NPT) as a declared nuclear weapons state. As a matter of international law, there was nothing illegal about following that path. China and France had just done that a few years earlier. But Israel's strategic situation and unique relationship with the United States made it different, fundamentally so, from all previous proliferators. Because of these considerations, Israel's political leadership was profoundly hesitant about the degree of its nuclear intentions and commitments.

One thing was clear: Prime Minister Eshkol categorically ruled out conducting a nuclear test on political grounds. "Do you think that the world would congratulate us for our achievement?" he used to ask sarcastically those people around him who entertained the idea of a test.[39] He had good reasons to reject a test outright.

First, Eshkol knew that a nuclear test would be a blatant violation of his pledge that Israel would not be the first to introduce nuclear weapons to the Middle East. While Israel left the exact meaning of "nuclear introduction" vague, it was plain that that commitment was totally incompatible with nuclear tests.[40] Second, Eshkol was aware that the superpowers were leading

---

vices as early as 1966. John W. Finney, "Israel said to buy French missiles," *New York Times* (January 7, 1966). As early as 1970 the CIA cited "Jericho" as the name of the Israeli missile (its original French version was known as MD-620 or MD-660), claiming that France tested eleven such missiles for Israel. Hedrick Smith, "U.S. assumes the Israelis have A-bomb or its parts," *New York Times* (July 18, 1970).

[39] The quote came out in an Interview with Zvi Dinstein, Eshkol's trusted aide and the Deputy Minister of Defense in 1966–7. Subsequently, a test site was chosen, and certain preparations were made to make it possible for Israel to conduct a test, following a political decision, in a relatively short notice.

[40] Indeed, in his speech in the Knesset in May 1966 Eshkol confirmed that interpretation when he stated plainly that Israel had no nuclear weapons.

negotiations on a global treaty aimed at preventing the proliferation of nuclear weapons. Conducting a nuclear test would not only be a catastrophe for American-Israeli bilateral relations—relations in which Eshkol had placed so much political capital to cultivate—but also an act of defiance against the will of the entire world community. Third, the prohibition against nuclear testing meant more than merely a commitment to the United States. It was also a domestic promise that Eshkol willingly granted to his close political allies, the ministers Galili and Allon, the leaders of Achdut Ha'avodah. For them, the "non-introduction" commitment was not viewed as a concession to the United States but rather as a genuine Israeli strategic interest. They firmly opposed nuclear testing. Finally, of course, there was the Egyptian factor. Eshkol knew that an Israeli test would be catastrophic from a regional point of view, in particular in terms of the Egyptian response. It was evident that an Israeli test would likely provoke Egypt and possibly the Soviets into hostilities toward Israel, possibly provoking an all-out war. One way or another, it was evident that Israel had no option to conduct a test.

Putting the test issue aside, Israel needed to figure out its response to a set of complicated and interrelated issues involving the future of its nuclear project. Simply put, Israel had to define—in operational and political terms—the meaning of its nuclear commitment:

- What should be the strategic role of the nation's nuclear program for the post-R&D period?
- What should be Israel's real interest: to possess nuclear weapons secretly or to obtain the political assets that nuclear weapons could buy?
- How could Israel extract deterrence value from secret weapons?
- Could the nuclear program be used as a bargaining chip in a larger political deal, either with the United States or with Egypt? Should Israel pursue such a bargain?
- What did Israel actually mean when it committed itself to non-introduction of nuclear weapons? Was this a genuine Israeli interest or just a convenient formula to deflect American pressure?
- How should Israel operationalize its nuclear option? Should it include weaponization or even deployment?
- Should Israel move the Dimona nuclear infrastructure to a mode of production? Would that be compatible with the declaratory stance of non-introduction?
- What should be the future of the missile project?

In 1966–7 Israel had no clear-cut answers to these questions. Ben Gurion had left those long-term issues unsettled for years, even untouched. Now the project was fast approaching the threshold point, and they had to be addressed. The challenge was nothing less than squaring the circle, i.e. finding the right balance between the two fundamentally opposite horns of Israel's nuclear dilemma, between technological resolve and political caution. Israel had to design a nuclear posture that would preserve a "nuclear option" but in a manner that would still be compatible with its "non-introduction" commitment. In a way, Israel was facing a moment of truth for the nuclear project.

For the cause of resolve, there were the project's leaders who naturally pushed for moving forward. For them, it was nearly inconceivable to bring the project to a pause at such a critical junction. The very ethos of the project, as they understood it, was that a nuclear option meant an *operational* capability available for the existential moment of last resort. Freezing the project in a non-deployable mode was unthinkable. Israel had to retain a real nuclear option, not something virtual and amorphous. They felt that it would be impossible to stop a moving train.

On the side of caution, however, there was Israel's commitment not to be the first to introduce nuclear weapons to the region. This formula had been used first in 1962 by Ben Gurion, and a year later by Peres who used it as a response to a query by President John F. Kennedy, but it was Prime Minister Eshkol himself, in a memorandum of understanding that he signed with the United States in March 1965, who made it a pillar in US-Israeli security relations.[41] Israel left the exact meaning of its commitment vague—it never made the operational meaning of this pledge explicit and the United States did not press for clarifications—but it was logically understood that "non-introduction" meant, at the minimum, the non-testing and non-possession of nuclear weapons. Indeed, in his speech in the Knesset in May 1966 Eshkol stated plainly that Israel had no nuclear weapons and would not be the first to introduce them to the region.[42]

But Israel's commitment to nuclear caution was more than merely a commitment to the United States. It was anchored in domestic and strategic considerations that went far beyond American-Israeli relations. In the eyes of Eshkol's close political allies, Galili and Allon, this commitment was a renewed formulation of what Ben Gurion had already pledged to them in

[41] Cohen, *Israel and the Bomb*, pp. 207–15.

[42] Ibid., pp. 231–5.

1962, and as such it manifested an Israeli domestic commitment about the nation's nuclear program. It constituted a delicate and tacit Israeli political consensus that allowed keeping the nuclear program running quietly in the background.

While the nuclear project was quickly moving toward the nuclear threshold, politically Prime Minister Eshkol was unsure and unclear about where the project was heading. There was an aura of anxiety. It appears that by early 1967 Eshkol was more open than ever to creative ideas that would allow him to avoid crossing the threshold, or possibly just to slow Israel's approach toward the threshold. There are clear indications that Eshkol was hesitant, ambivalent, and uncertain as to where the nuclear program should go once the R&D stage was completed.

But it was the crisis prior to the 1967 Six Day War that fatefully overwhelmed his hesitancy. New and unexpected developments that occurred in those fateful days changed the entire situation and probably determined a great deal of the history to come. Without making any political resolution regarding the nuclear dilemma—no such resolution was forthcoming anytime soon—the very dynamics of the crisis generated new facts, new attitudes, and a new frame of mind.

As Israel was facing the worst crisis in its history—the massing of Egyptian troops in the Sinai Peninsula and the possibility of an imminent surprise aerial attack—the nuclear project became, almost overnight, the nation's ultimate insurance policy. At a time when Israel was setting up temporary burial sites for thousands of soldiers, it was unthinkable for the leaders of the nuclear project to sit idly by. Their philosophy was that if the capability could be available, it must be available. And surely at that time Prime Minister Eshkol was not in a position to stop them. He, for his part, authorized special activity to accelerate nuclear progress.

In those days running up to the war, Israel did something it had never done before. In an intensive crash effort Israel teams improvised the assembly of the nation's first nuclear explosive devices. As Israeli scientists and technicians "tickled the dragon" while they assembled the first nuclear core for those devices, only a few of them were even aware that there was a military contingency plan in the works. In response to the worst of scenarios, and in particular to fears about the failure of the Israeli Air Force and the extensive use of Egyptian chemical weapons against Israeli cities, some preliminary plans were drawn up to show that those devices could be used in a manner that would demonstrate capability but fall short of actual military use ("use in anger").

The idea behind it was indicative of Israel's level of anxiety in those days. If all else failed, and Israel's existence was in peril, Israel would still have a final doomsday capability. Despite these threats, though, the final step in the assembly process, arming the devices, did not take place.[43]

These were extremely dramatic moments for those who were involved, particularly for the project's leaders. In their mind, the assembly of those devices meant an extraordinary historical moment in the life of the project and the nation as a whole. It was seen as the moment that Israel became a nuclear power.[44] From their perspective, it was also an irreversible moment.

[43] During that intense crisis, Shimon Peres, the former leader of the nuclear project and then an opposition party leader, made the suggestion to test the devices in a semi-public fashion as a deterrent move designed to avert war. Peres, like Ben Gurion, was looking for any possibility to avert war, fearing that war would be disastrous for Israel. It is impossible, of course, to predict what would have been the political, military and psychological impact of such a spectacular move, and whether it would have actually helped to "resolve" the crisis in a peaceful way. Surely a nuclear test would have introduced a most dramatic new issue into a complex political environment that had been already highly explosive. While clearly Peres considered the crisis situation as Israel's best opportunity to get international legitimacy and sympathy for the bomb, it could be argued equally that given the intense crisis situation—including concerns over the reaction of the Soviet Union—the big powers would have demanded forcefully that Egypt should withdraw its forces from the Sinai and open the straits while Israel should give up the Bomb and accept international safeguards on its nuclear program. Despite the high risks involved, Peres' idea made some sense. After all, what else had the Bomb been built for if it were not to avert possible calamity in a war of survival? One should expect the IAF to win the air battles but you could not know in advance that there would be complete surprise and total victory in the air war with so few casualties; it was not fully assured. If the war had started and gone badly—if, say, the Egyptians had attacked early and wiped out much of the Israeli Air Force on the ground, following this by a sweep into the south—any test would have looked like and in effect would have been use-in-war. There might have been a second Cuban missile crisis with Russia and US building up nuclear forces in the area, but that would have been a small price to pay for Israel. Of course Peres had a vested interest in making the nuclear weapons program public because it would have made him look like the man whose project saved the nation. As it turned out winning the war and keeping the Bomb quiet was the best of all possible worlds for Israel, but I'm not sure it would have looked that way in May 1967.

[44] One of them recalls, as he told me decades later, that he proposed after the war to seize the moment and to conduct a test with one of those cores. His proposal was never seriously considered. "It was a total taboo to them," he recalled years later. This

They could no longer conceive of a future Israeli prime minister who would give up this capability for any reason, except perhaps a real peace.

For all we know, Prime Minister Eshkol made no major formal decision on the nuclear question. On the contrary, as already noted, prior to the war he was hesitant, ambivalent and unsure. Yet, even without making a decision, the balance between the forces of resolve and caution tipped in those days in 1967 in a fundamental way toward resolve. While Eshkol may have been leaning toward the side of caution, the post-R&D dilemma was overwhelmed by events. Judging by the facts on the ground, Israel, as it approached the nuclear threshold, appeared to have made a commitment to possess nuclear weapons.

The NPT dilemma

While the significance of the new facts on the ground had to be reckoned with, Israel still had to respond in a comprehensive manner to the many outstanding issues that constituted its post-R&D nuclear dilemma. Israel still needed to define in operational and political terms the meaning of its nuclear commitment. Eshkol's task was further complicated due to the political and personnel changes that the Six Days War brought about. Until the war Eshkol held the dual portfolio of Prime Minister and Minister of Defense, and as such, he was the sole boss of the nuclear project. After the war, with the appointment of Moshe Dayan as Minister of Defense, the responsibility for certain aspects of the nuclear project had to be shared between them.[45] Furthermore, Zvi Dinstein, Eshkol's right-hand man on the nuclear project, was no longer the deputy at the Ministry of Defense. His executive responsibility was passed to Dayan's top civilian aide, former Chief of Staff Zvi Zur, who was friendlier to the needs of the nuclear project than Dinstein. Moreover, the

---

shows the strength of nuclear caution at the political level, but one can only speculate how Israel would have been had that proposal was accepted.

[45] With the appointment of Dayan as Minister of Defense, Prime Minister Eshkol asked the former Chief of Staff Yigal Yadin to write a document, referred to as the "constitution," that defines the division of labor between the Prime Minister and the Minister of Defense. Theoretically, the formal executive authority over the nuclear matters remained in the hands of the Prime Minister, but in practice much of the budgetary responsibility was with the Minister of Defense. See Yehudah Ben Meir, *Civil-Military Relations in Israel* (New York: Columbia University Press, 1995), p. 101.

cabinet was now reinforced by a Minister of Defense who was himself a nuclear advocate.[46]

However, it appears that neither Eshkol nor Dayan was interested in opening up the nuclear issue for a cabinet debate, and they surely were not interested in a repeat of the 1962 debate.[47] They simply were not interested in politicizing the nuclear issue. In any case, Eshkol's plate was full after the war, and by mid-1968 his health was fast deteriorating. Eshkol died in office in February 1969 not knowing what Israel's nuclear future would look like. This task was passed to Israel's new leader, Prime Minister Golda Meir.

But perhaps the most important development was that the post-1967 war nuclear dilemma was reframed and dominated now by the advent of one major factor. A new nuclear dilemma had arisen in the form of Israel's stance on the nuclear Non-Proliferation Treaty. Israel was required to make up its mind where it stood on the matter of the NPT. Deciding this would inevitably determine a great deal on all other outstanding nuclear issues.

It turns out that the completion of the NPT and Israel's completion of its R&D phase took place virtually in historical parallel, as if competing with each other. It took each development nearly a decade to ripen; the completion of each meant an end of an era. But most significantly, each represented historical trends that were in deep contradiction with each other. But until late 1968 very few in Israel (and in the United States) understood the depth of that conflict.[48]

It took quite a long time for Israel to resolve the NPT dilemma. In the meantime, the nuclear program was muddling through, drifting slowly toward

[46] The cabinet was divided between Minister of Defense Moshe Dayan, a strong advocate of nuclear resolve, and the security oriented ministers of Achdut Ha'avodah, Deputy Prime Minister Yigal Allon and Minister without Portfolio Galili. Prime Minister Eshkol, who since mid-1968 was increasingly troubled by medical issues, did not have the charisma and authority that Ben Gurion had on this issue.

[47] There are hints that sometime in the second half of 1968 Eshkol passed the issue of the NPT to a small ad-hoc ministerial committee.

[48] Not many people are aware today how important the Israeli case was for the evolu\ tion of the NPT. I elaborated on this issue elsewhere, but I should remind the reader that from an American perspective Israel was hovering between the irreversible proliferation of the past and the new non-proliferation prohibition movement of the future. It was on that divide that the new non-proliferation norm would enforce a new global order. Since the mid-1960s it was hoped that the NPT would be able seal Israel's nuclear status on the non-nuclear side, even if Israel was not allowed to cross the threshold. Avner Cohen, "Before the Non-Proliferation Norm: Israel's Nuclear History," *Current History*, Vol.104, No. 681 (2005): pp. 169–75.

the threshold point, while no top-down political decision was formally made. In response to the vacuum, various governmental agencies—civilian and military—made their own strategic assumptions about the future, in bureaucratic isolation and in secrecy from other agencies.[49] Nuclear ambiguity was the name of the game in those days, with respect not only to the United States but to the bureaucracy itself as well.

By spring 1968 the conventional wisdom, both in the US and in Israel, was that Israel would eventually join the NPT.[50] Even without access to Israeli internal deliberations one can reconstruct the logic behind the reasoning. It was generally assumed that obtaining Israel's signature to the NPT was more important to the United States than it was to Israel to stay outside the treaty. For Israel, to reject the NPT was viewed as a defiant act on an issue that was important for the US's global interests. It must be remembered that in 1968 Israel was anxiously anticipating a fundamental transformation in its security relations with the US; America was about to replace France as Israel's prime defense supplier. The conventional wisdom was that, in return for the Phantom sale that Eshkol had officially requested in his visit in January, Israel would sign the NPT.[51]

Equally important was the view that signing the NPT was compatible with previous Israeli nuclear commitments, in particular the "non-introduction" pledge. It was thought that signing the NPT added little, perhaps nothing, to what Israel had already committed itself not to do. After all, Israel had already

[49] Possibly seven bureaucracies were involved, at one level or another, in some aspects of strategic thinking that involved the future of the nuclear project. There was very little coordination in their thinking, often each body treated its own thinking as highly classified and not to be shared with the other bodies. Often, there were major disconnects between those bureaucracies. Here is a short list of those bureaucratic bodies: the Prime Minister's office (reporting to Yaakov Herzog), the Israel Atomic Energy Commission (reporting to Professor Dostrovsky), the Ministry of Defense (reporting until May 1967 to Zevi Dinstein and afterwards to Zvi Zur), the Foreign Ministry (reporting to foreign minister Abba Eban), the Israel Defense Forces (IDF) (reporting to Chief of Staff Rabin), the Israeli Air Force (reporting to the commander of the Air Force General Hod and Brigadier General Peled), and Mossad (reporting to Chief of Mossad Meir Amit).

[50] Cohen, *Israel and the Bomb*, pp. 299–300.

[51] This was exactly the argument that Deputy Secretary Paul Warnke raised in linking the NPT with the Phantoms sale during the tough negotiations he had with Ambassador Rabin. Cohen, *Israel and the Bomb*.

committed itself not to be a nuclear power. Moreover, signing the NPT would still allow Israel to keep a posture of advanced "nuclear pregnancy."[52]

But those early expectations all proved wrong. The NPT was presented for signature on July 1, 1968. Sixty-five nations signed it on that day, but Israel was not among them.[53] The more Israel studied the legal and political implications of the NPT, the more doubts it had. By the fall of 1968 Israel raised in public various concerns about the applicability of the NPT to its security situation, in particular the concern that the NPT lacked guarantees against aggression by a nuclear power (such as the Soviet Union) against a non-nuclear state. Despite these public concerns, though, Israel's real problem with the NPT could not be spelled out in public.[54] Evidently, the NPT issue forced the Israeli government to take a hard look on the entire nuclear issue, something that it had been delaying for a long time.

Israel must have recognized that signing the NPT meant a solemn legal obligation not to acquire nuclear weapons. Probably there were those who warned that signing the NPT was utterly incompatible with Israel's newly acquired nuclear capabilities. Taking the NPT seriously would have required, legally speaking, rolling back those capabilities. But Israel could not convey this to the United States—doing so would have meant that Israel had been manipulating American visits to Dimona for years—and so Israel formally maintained that the NPT was still "under active study." As I noted elsewhere, the NPT issue came to a full-blown showdown during the negotiation of the

[52] If this analysis is true, it highlights how problematic was the way Israel dealt with its own nuclear situation. Due to a combination of secrecy, ambiguity and inhibition, senior Israelis—surely officials at the Foreign Ministry, but possibly even Prime Minister Eshkol himself—did not realize that the NPT was inconsistent with Israel's nuclear capabilities, and if Israel had signed the Treaty it would have been legally required to roll back those capabilities. Those Israeli officials who expressed the view that Israel would likely sign the NPT had very little understanding of the NPT and its far-reaching legal and political implications. It also appears that at that point there was still little coordination among the various aspects of the Israeli bureaucracy that dealt with the nuclear issue.

[53] Israeli sources still maintained that Israel would sign the NPT soon. "Israel conducts consultations of the Nuclear Treaty", *Ha'aretz* (in Hebrew) (July 2, 1968).

[54] "Israel will set its policy after the Non-Nuclear State Conference", *Ha'aretz* (in Hebrew) (August 30, 1968). By 1978 all three co-custodians of the NPT made statements that pledged negative security assurances. Thomas Graham and Leonor Tomero, "Obligations for Us All: NATO and Negative Security Assurances," *Disarmament Diplomacy*, Vol. 49 (2000).

Phantoms sale, with Israeli Ambassador Yitzhak Rabin refusing to accept any linkage between signing the NPT and receiving delivery of the Phantoms. Now we know that it was Rabin himself who urged the Israeli cabinet not to accept American demands on the matter. In the end, following presidential intervention, the deal was signed without any NPT linkage and the NPT issue remained outstanding.

By the summer of 1969 it became evident that in order for Israel to say "no" to the NPT it had to first quietly work out new understandings with the United States, and this had to be done at the highest level. That task was passed to Golda Meir, Israel's new Prime Minister, in her first meeting with President Richard Nixon in September 1969. New bilateral nuclear understandings were agreed upon by the two leaders: the United States would tolerate Israel's nuclear capability as long as Israel would handle it with the utmost caution. That meant practically that Israel was committed to three major constraints: not to conduct a nuclear test, not to change its declaratory policy (that is, not to declare its nuclear status), and not to threaten any country with its capability. In brief, the US looked the other way at Israel's nuclear weapons, but they had to be hidden in the basement.

In September 1969 Prime Minister Meir laid the groundwork for Israel's negative decision on the NPT, but she still promised a formal reply after her new cabinet had discussed the issue. It took four additional months until Ambassador Rabin informed National Security Advisor Henry Kissinger in February 1970 that Israel had decided not to sign the NPT. Rabin provided no further explanations, he just asked Kissinger to notify President Nixon of Israel's decision. Kissinger did not ask Rabin for any explanation either, nor did he even express regret for the record. He only told Rabin that he would inform the President of Israel's decision. Nixon probably was not surprised by Israel's decision. And that was that.

Sometime around that time, during the second half of 1969 or early 1970, there was a top Israeli leadership meeting on the future of the nuclear project at Dimona. Among those who attended the meeting were Prime Minister Golda Meir, Minister of Defense Moshe Dayan, Deputy Prime Minister Yigal Allon, Minister Yisrael Galili and the leaders of the nuclear project. No documents are available about the meeting, but according to a testimony of one participant, Professor Yuval Ne'eman, who was then a member of the IAEC, Minister Dayan dominated the meeting. According to Ne'eman, Dayan seemed to discuss issues that were quite irrelevant to the subject of the meeting—citing recent intelligence reports about torture in the Arab world, in particular graphic accounts of how Arab leaders tortured their political opponents. A few minutes into his talk, according to Ne'eman, Meir interrupted him by saying, "Enough with this. I made up my mind. We will proceed."[55] In Ne'eman's view this tactic was brilliant; it was Dayan's way of reminding his colleagues the kind of neighborhood in which Israel lived. While Ministers Allon and Galili were willing to put some unspecified limits on the nuclear program, possibly on the production of fissile material, Meir was persuaded that Israel must proceed with the program full blast. According to Ne'eman, this was a key political decision that preceded Israel's final negative decision on the NPT.[56]

All of those nuclear decisions in 1969–70 were elements in one major policy decision that the Meir government made. The understandings between Nixon and Meir made it possible for Israel to formally decide not to sign the NPT and to move into a real nuclear deployment posture. Moreover, the United States went along with this. The understandings between Nixon and Meir meant a great deal to Israel in that it determined what Israel could and could not do in the nuclear field. Because secrecy was so essential for the deal, it left Israel with much freedom in executing its own share of the deal. As long as those activities remained invisible, Israel was left free to pursue it on its own. A decade of American pressure was replaced now by Israeli self-constraint and secrecy.

After years of hesitancy and inhibition, Israel had finally made up its mind on its post-R&D nuclear dilemma. It had chosen resolve over caution. This policy decision was among Israel's most fateful decisions it ever made, perhaps equal to Ben Gurion's initiation decision in 1957. It left a profound legacy for it shaped the basic parameters of Israel's nuclear policy and conduct for decades to come. In the next few years, under the careful leadership of IAEC Director General Shalheveth Freier, those political decisions were translated into a well thought-out formulation of Israel's nuclear posture of opacity. Much has changed since then, of course, but that blueprint is still with us today. It allowed Israel to move its nuclear posture forward, although with much caution, into production and into deployment mode. An element of Israel's caution manifested itself in a deeply held and widespread societal taboo that holds to this day.

I have discussed some of the bilateral and international aspects of that legacy elsewhere, but here I want to add a few words on the domestic aspect of

[55] Yavul Ne'eman, many conversations with author, 1995–6.
[56] Ibid.

that legacy. The opacity deal of 1969 put to rest another Israeli nuclear dilemma regarding domestic disclosure of the nation's nuclear status. That dilemma after the opacity deal became moot. The 1969 deal committed Israel to a policy of non-acknowledgment, both internationally and domestically. Thus, in response to those decisions, the rules of the game as to how the censor supervised the Israeli discourse on nuclear matters were rewritten. It was decided that the Israeli press would be allowed to refer to Israel's nuclear capability in terms of a "nuclear option" or "nuclear capability," keeping the words "bomb" or nuclear weapons" out of the public discourse. This is still the practice of the Israeli military censor today.

All subsequent Israeli prime ministers to this day have endorsed the legacy of the Nixon-Meir deal. Resolve and caution have come to co-exist side by side, both wrapped in the shroud of nuclear opacity. And this—the shroud of opacity—has made Israel distinct as a reluctant proliferator.

The use dilemma

A nation that develops nuclear weapons must inevitably grapple with the use dilemma: the circumstances in which nuclear weapons could be used. In order to design a militarily meaningful posture, there needs to be thinking about the military utility of those weapons. Even if the thinking about the purpose of the nuclear program is primarily political—projecting existential deterrence and overall strength—weaponization necessitates thinking about military use.

The very idea of deterrence is built upon credibility, and credibility is based on the feasibility of use. If the products of the nuclear program are in non-weaponized form, their deterrent value may be diminished because they may be perceived as less credible. On the other hand, leaving the products of the nuclear program presumed but unacknowledged could still generate an effective deterrent value; it all depends on perception.

In the mid to late-1960s Israeli defense intellectuals were asked to grapple systematically with these issues. Initially, those discussions were made in quasi-theoretical fashion: Were Israel to develop and acquire nuclear weapons, under what circumstances they could be used? What kind of nuclear weapons would best serve Israel's needs? What kind of arsenal might Israel need? It should be noted that it was not the IDF that initiated this brainstorming—it was done on Eshkol and his deputy at the Ministry of Defense, Zvi Dinstein's initiative and in coordination with Chief of Staff Yitzhak Rabin; rather, IDF senior personnel were invited to contribute their own intellectual insight, along with the insight of others, on nuclear matters. Apparently those discussions advocated the view that Israel's commitment to nuclear resolve must, at a minimum, reach the level of full or near-full weaponization. It was argued that if Israel took the pains to develop a nuclear capability for the unlikely moments of existential threats and to project deterrence, that capability must be operational, not merely virtual. Such a view was highly influential in the Israeli government, under Golda Meir, when it made its final decisions on the nuclear issue in the 1969–70 period.

These early deliberations produced the first articulation of the underlying concept of Israel's use doctrine. The idea was to define certain existential "red lines" that, if crossed, could trigger the use of nuclear weapons. Those "red lines" gave operational meaning to the somewhat vague idea of "last resort." Specifically, four distinct scenarios—all of them were considered highly theoretical and extremely unlikely—were identified that could invoke such use: (1) a successful Arab military penetration into populated areas within Israel's 1949 borders; (2) the destruction of the IAF; (3) the exposure of Israeli centers of population to massive, devastating and continuous air attacks, or to major chemical or biological attacks; (4) the use of nuclear weapons against Israeli territory. Each of these scenarios was defined as an existential threat to the State of Israel against which the nation would be justified to defend itself by no other means than with atomic weapons. The use of atomic weapons under such circumstances would be politically, militarily and morally justified.[57]

From the very start, however, it was evident that finding a sensible use of nuclear weapons in the context of the Arab-Israeli conflict would be highly problematic. Analysis of the very notion of "last resort," especially under circumstances in which the capability is unacknowledged and unintroduced, highlighted a certain paradox. It became apparent that to use a nuclear bomb *after* a massive Arab army had penetrated Israel's borders could be too late to be militarily effective, perhaps even impossible because of the proximity of Israeli troops (or citizens). Conversely, using nuclear weapons to preempt Arab army troops on their way to the border was deemed too early to justify use—that is, politically unacceptable. Israeli strategists encountered a problem quite similar to the one faced by NATO throughout much of the Cold War: the difficulty of defining the proper moment—militarily and politically—at which nuclear weapons could be used effectively to stop a conventionally superior enemy attack.[58]

[57] Cohen, *Israel and the Bomb*, chapter 12.

[58] Yuval Ne'eman, "Israel and Nuclear Deterrence" (in Hebrew), *Ma'archot*, Vol. 308 (1986): pp. 19–21.

Another realization that emerged in those discussions, in part in response to this dilemma, was that it would be inconceivable for Israel to resort to nuclear use in war without an explicit prior warning. It was agreed Israel's first nuclear detonation must be in the form of a demonstration for deterrence purposes. After all, deterrence works only if the capability and the willingness to use are clearly conveyed to the other side. But how can a state deter if it does not acknowledge that it possesses the bomb in the first place? How is it possible to establish deterrence effects under a strategic situation of nuclear ambiguity? This is even more complicated when a nation is committed to a political posture of non-introduction, which logically appears to imply no first use. A nuclear demonstration has two important values: it is the equivalent to issuing a deterrent statement and it means an act of acknowledgement; a demonstration can be conducted earlier than the actual crossing of those "red lines."

Given that the capability became real and available, there was a need to contemplate the circumstances under which it could be used, or, more accurately, the circumstances under which decision makers would be willing to consider the nuclear option. It was clear from the start that the use of atomic weapons was totally incompatible with the IDF operational plans, which were all based on aerial preemption to be followed by an Israeli armor attack deep into the Sinai. It turned out that efforts to rationalize scenarios of atomic use illustrated the strangeness involved in such thinking about the unthinkable. All involved various doomsday scenarios that assumed a colossal failure by the IDF in its preemption and a decisive strategic surprise by Egypt—for example, massive use of missiles tipped with chemical warheads against Israeli cities. Still, all contingency plans were based on using the capability in a political fashion first through a demonstration.

In reality, none of these scenarios materialized. The military doctrine of conventional preemption won the day. Israel's conventional success in 1967 demonstrated that Allon and Galili were right in the 1962 debate. One important lesson could have been that the nuclear issue is primarily about politics and psychology, not about military utility. Israel was strong enough conventionally to defeat three Arab armies in six days. The nuclear issue had proved totally unnecessary, a psychological extra for very unrealistic doomsday scenarios. Furthermore, Israel's improvised nuclear deployment extracted no deterrence value whatsoever. Shimon Peres, who in 1967 was outside the government, proposed the idea that Israel should take advantage of Egypt's massing of troops in order to introduce its nuclear capability by conducting a nuclear test.[59] His line of thinking must have been that the new circumstances that Egypt created opened a unique opportunity for Israel to introduce and legitimize its nuclear status. Peres' call for nuclear resolve was ignored by the Eshkol government. Nuclear caution won the day and a legacy stressing the improbability of nuclear use was planted.

Another strategic lesson that Israel probably learned from the 1967 nuclear incident was the difficulty of extracting deterrence value from a situation in which the very capability itself was unacknowledged. The credibility of nuclear deterrence under a non-acknowledgement posture depends on perceptions, and perceptions can be manipulated even if they are not "introduced" officially. It has been suggested that one could think of nuclear deterrence in terms of an epistemic ladder that starts with unconfirmed rumors and speculations and ends with an official declaration accompanied by a test. In this sense Israel stands on, or just above, the first rung of that ladder. Israel's nuclear deterrence under a non-acknowledgement posture has rested on the presumption—periodically supported by credible leaks—that Israel actually had the bomb and, under conditions of existential threat, would use it. It was also proposed after the 1967 war that in the case of an emergency Israel needed to be ready to move quickly up the deterrence ladder, which meant that Israel needed to have the technical capabilities to demonstrate its nuclear capability at very short notice. In theory, those views were articulated in various advisory groups, and it was assumed that the final decision was up to the political leadership. In reality, those views generated a drift toward nuclearization even when the political level issued no formal decision.[60]

Perhaps the most dramatic moment in Israel's nuclear history took place during the 1973 war. Unlike the situation in 1967, Israel approached the brink of defeat. Its defense lines on both fronts, the Suez Canal and the Golan Heights, were overwhelmed by huge advancing columns of armor and infantry, backed up by hundreds of pieces of artillery. Hundreds of Israeli soldiers died in the early hours of the war as many regular army units were disseminated. It was the closest Israel had faced to a colossal collapse. Defense Minister Moshe Dayan, who was in a gloomy state of mind and murmured about "the demise of the Third Kingdom, "was reported as proposing ideas involving Israel's ulti-

[59] Cohen, *Israel and the Bomb*.

[60] Avner Cohen, "Nuclear Arms in Crisis under Secrecy: Israeli 1967 and 1973 Wars" in Peter Lavoy and Scott Sagan (eds), *Planning the Unthinkable: How New Powers Will Use Nuclear, Biological and Chemical Weapons* (Ithaca: Cornell University Press, 2000), pp. 104–24.

mate weapons. Dayan was fearful that Israel was approaching a point of 'last resort'; evidently, he wanted the United States to take notice that Israel had reached that point."[61] No record has surfaced to this day of what exactly Dayan proposed, and what activities took place on the ground, but rumors referred to various doomsday scenarios that Dayan could have invoked.

This set the stage for a moment that defined the other aspect of Golda Meir's nuclear legacy. She refused to agree to nuclear use. Ultimately, it was she, the Prime Minister, who made the real decisions on the use of those ultimate weapons. She saw nuclear weapons as political weapons, not military weapons. Like John F. Kennedy a decade earlier, Golda Meir stared into the nuclear abyss but refused to concede to it. Her nuclear conduct demonstrated that Israel was a responsible and cautious nuclear custodian. It highlighted the recognition that nuclear weapons were not like any other weapons humanity had ever invented; that under virtually any circumstances they must not be used. The side of caution had won. This legacy is as relevant today as it was then.

More than four decades have passed, including three major wars, since Israel started to grapple with thinking the unthinkable, but little has changed in the fundamental Israeli thinking on these matters. If anything, the experience of the wars deepened the basic Israeli outlook that nuclear weapons may have symbolic and political value but they lack genuine military value and should not be recognized as military weapons systems. Generations of Israeli defense intellectuals rediscovered one and the same lesson: how difficult, indeed nearly impossible, it is to construe real military scenarios that could call for—let alone justify—the use of nuclear weapons. As Israel recognized in the 1967 war, nuclear thinking stands at odds with the principles of Israel's conventional warfare strategy. Even in the 1973 war, the war that brought Israel the closest to the brink, it would have been extremely difficult to construe a scenario in which nuclear weapons could be used sensibly. Israel has no military reason to "introduce" nuclear weapons and it has many political reasons not to be the first to introduce those weapons.

The central political role of nuclear weapons is to spread the notion that calls for the "destruction of Israel" are irrational. If those "last resort" scenarios made little sense in the strategic context of the mid-late 1960s, they make even less sense as Israel's conventional superiority has only been compounded vis-à-vis the Arabs. If under the conditions of 1967 it was nearly impossible to conceive of "last resort" situations in which Israel could consider using nuclear weapons, such reservations must have deepened and strengthened ever since.

[61] Ibid.

Just as Israeli planners had realized initially the enormous difficulty of conceptualizing situations requiring nuclear use in any militarily meaningful or rational way, so did they learn this lesson on all three concrete occasions when Israel seemed on the brink of devastating military defeat. The validity of much of this thinking regarding the utility and use of nuclear weapons has remained little changed. All of these cases indicate that short of a nuclear attack, it is inconceivable that Israel would use nuclear weapons to defend itself against existential threats. The legacy, then, is that Israel nuclear weapons are truly unusable.

Subsequent dilemmas: advanced weaponry and posture design

New types of nuclear decisions are called for as a country moves from initiation to maturity. Those subsequent nuclear dilemmas involve posture and arsenal design: how big and how advanced the nuclear commitment should be. We know that each of the first five nuclear weapons states moved quickly to develop advanced nuclear weaponry, in particular thermonuclear weapons, within a decade or so. I will try here to sketch the ways in which Israel has dealt with advanced nuclear weapons. It must however be kept in mind that the unacknowledged nature of Israel's weapons program in general renders such a sketch no more than an educated guess.

There are indications that after the 1973 war Israel committed itself to more advanced nuclear weaponry. It is believed that Israel's current nuclear arsenal is qualitatively advanced and consists primarily of two-stage thermonuclear weapons. But what is even more interesting is Israel's negative decision regarding tactical battlefield nuclear weapons. Unlike all declared nuclear weapons states, which developed, manufactured and deployed tactical nuclear weapons (except China), Israel made a different decision. Israel apparently decided against the production and deployment of tactical nuclear weapons for the battlefield (even though it completed the R&D phase necessary for their development). The reasons behind that decision were many—political, strategic and operational—but apparently they all manifested a concern over blurring the distinction between nuclear and conventional weapons and some operational concerns that the possession of tactical nuclear weapons would undermine some of the fundamental aspects of nuclear opacity. Given the traumatic experience of 1973, this was a bold strategic decision.

Finally, there is uncertainty about the size of the Israeli nuclear arsenal.[62] Following the Vanunu revelations it was commonplace to hear claims that

[62] Cohen, *Israel and the Bomb*, p. 12.

Israel may possess 100 to 200 nuclear weapons. Seymour Hersh even alleged that Israel might have 300 nuclear weapons or more, both tactical and strategic.[63] But more contemporary estimates, based on leaks attributed to sources within the US intelligence community, indicate a much smaller arsenal, perhaps in the neighborhood of 100 weapons, possibly fewer.[64]

If these more conservative claims are true, the development of Israel's nuclear posture was guided by strategic considerations derived from Israel's nuclear doctrine, and not from technological or bureaucratic inertia as some feared in the wake of the Vanunu revelations in 1986. It suggests that Israel dealt with its subsequent nuclear dilemmas in a spirit of careful calibration between resolve and caution. It is consistent with Israel's other strategic decision in the mid-late 1970s to develop a longer-range and more accurate nuclear-tipped ballistic missile, the Jericho II, to replace the outdated Jericho I. Israel reportedly tested the Jericho II in the late 1980s and deployed it in the early 1990s. Those decisions underscore Israel's commitment to nuclear deterrence.

If it is true that Israel decided not to produce and deploy tactical nuclear weapons, and also kept the size of its arsenal relatively small, this indicates that, in contrast to the historical trend of most other nuclear nations, Israel treats its nuclear arsenal in existential terms. If so, these decisions are aligned with previous Israeli nuclear decisions to maintain the conventional orientation of the Israeli army as well as the related requirements intrinsic to the regime of opacity.

63 Seymour Hersh, *The Samson Option: Israel's Nuclear Arsenal and American Foreign Policy* (New York: Random House, 1991).

64 Mark Fitzpatrick (ed.), *Nuclear Programmes in the Middle East: In the Shadow of Iran* (London: International Institute for Strategic Studies, 2008), p. 133; "Revealed: the secrets of Israel's nuclear wrsenals," *Sunday Times* (October 5, 1986); Frank Barnaby, *The Invisible Bomb* (London: I.B. Tauris, 1989), p. 24; Seymour Hersh, *The Samson Option*, pp. 291, 312, 319; Anthony Cordesman, *Perilous Prospects: The Peace Process and the Arab-Israeli Military Balance* (Boulder: Westview Press, 1996), p. 234; Wisconsin Project on Nuclear Arms Control, "Israel's Plutonium Production," *The Risk Report*, Vol. 2, No. 4 (1996); Defense Intelligence Agency (DIA) estimate published in Rowan Scarborough, *Rumsfeld's War: The Untold Story of America's Anti-Terrorist Commander* (Washington, DC: Regnery Publishing, 2004), pp. 194–223; Kenneth Bower, "A Propensity for Conflict: Potential Scenarios and Outcomes of War in the Middle East," *Jane's Intelligence Review*, Special Report No. 14 (1997): pp. 14–5; Stockholm International Peace Research Institute (SIPRI), *SIPRI Yearbook 2007: Armaments, Disarmament and International Security* (Stockholm: 2007), p. 548.

From the early 1980s (and possibly earlier) the Israeli navy (with the support of other governmental agencies) promoted the idea that Israel should build a small fleet of modern conventional (diesel) submarines for "strategic purposes," an Israeli euphemism for a sea-launched nuclear capability. After complex negotiations in which a deal was almost signed with a German shipyard in early 1990, it was vetoed by General Ehud Barak, due to considerations of cost. In a way, though, his veto also reflected the prevailing sense of nuclear caution held by Israeli policymakers.

After the 1991 Gulf War, however, Israel's strategic picture changed fundamentally. Sometime after the war Israel decided to reverse its earlier decision and to establish a sea-based strategic arm. Israel accepted enthusiastically the decision of the German government to finance the purchase of two large diesel submarines and to share in the financing of the third (because of the role German industries played in the development of Iraq's unconventional weaponry). The strategic developments throughout the 1990s, in both Iraq and Iran, compounded by the failure of Western intelligence agencies to detect Iraq's nuclear program, were critical to the Israeli decision to boost its strategic capabilities. Nuclear resolve had taken the lead. Although details about the specific capabilities of those submarines are highly classified, it is presumed that that the new sea-based strategic arm has a second-strike (nuclear component). It is also understood that the most sensitive aspects of the submarine project, the cruise-missile technology that gives those diesel submarines dual-capable (nuclear and conventional) launching platforms, were developed and built in Israel. It is believed that the Israeli-made cruise missiles have the ability to hit strategic targets within a range of more than 900 miles.[65]

By July 2000, Israel had taken delivery of all three Dolphin-class submarines it had ordered in Germany. In doing so, Israel has moved toward acquiring a survivable second-strike nuclear capability. Israel was now on its way to restructure its nuclear forces into a triad form, like that of five declared nuclear weapons states. Initially a fleet of three submarines was believed to be the smallest fleet Israel needed to have a deployment at sea of one nuclear-armed submarine at all times. In 2006 Israel placed an order for two more submarines. A survivable deterrent fleet of five is now perceived as essential because of Israel's unique geopolitical and demographic vulnerability to nuclear attack, one that no potential nuclear enemy of Israel could ignore.

[65] Joseph Cirincione, Jon Wolfsthal and Miriam Rajkumar, *Deadly Arsenals: Nuclear, Biological and Chemical Threats* (Washington, DC: Carnegie International Press, 2005).

9

Iran's Nuclear Program

Kayhan Barzegar

Since the beginning of the Iranian nuclear standoff in 2002, the conflicting views between Iran and the West over the advance of Iran's nuclear program have taken the issue beyond national policies and related peaceful energy demands, and have connected it to regional and international security issues at large. The main challenge facing Iran and the West in the nuclear standoff is how to strike the appropriate balance between Iran's right to the peaceful use of nuclear energy, guaranteed under the NPT, and the possibility of weaponization.

The Iran-West standoff can be divided into three periods:

First, the period of ambiguity and distrust (2002–06): it was in 2002 that the West revealed concerns regarding the advance of the Iranian nuclear program. In this period, the two sides engaged in nuclear negotiations aimed at distinguishing each other's goals and wishes and reaching a solution. The turning point in this period is related to the voluntary suspension of uranium enrichment by Iran during the period 2004–06. In this respect, the EU's failure to solve the Iran nuclear question finally led to the suspension of negotiations and Iran's resumption of its enrichment process.

Second was the confrontation period (since 2006): in this period, the two sides attempted to carry out their respective objectives, a trend which has since continued as Iran insists on its right to enrichment. Iran's enrichment activity is centered on enriching uranium to the level of 3.5 percent, but uranium enrichment only reached 20 percent after the country was denied access to fuel for its Tehran Research Reactor (TRR). The West stepped up its political and economic pressures, bringing along the adoption of more sanctions, including Resolution 1929 of the UN Security Council. The main challenge during this period was the prevailing distrust between the two sides. Iran's concern was based on the assumption that the goal sought by the West, particularly the United States, through negotiations was the complete cessation of the Iranian nuclear program. For its part, the West was worried that Iran's ultimate goal was indeed access to nuclear weapons. Despite varying initiatives by both parties in the form of negotiating packages, neither side was able to bridge the divide and therefore negotiations yielded no results.

Third, and most recent, was the confrontation and interaction period, starting in October 2009 with the discussion regarding the swap of Iranian enriched uranium for the Tehran Research Reactor (TRR) in Turkey, and continuing to the present (March 2011). Iran has insisted that in addition to the nuclear issue, the negotiations should be widened and include other related strategic issues between the two sides. Iran's readiness to negotiate with the West is based on acquiring nuclear "capability" and reaching the point of no return. The leverage used by the two—Iran's continuing enrichment and the West's persisting adoption of international sanctions against Iran—compels both sides to interact with each other. At this stage, both sides are attempting to enhance their bargaining positions in order to maximize their advantages in the nuclear negotiations. As of January 2011, both sides continued to insist on their own positions at the Istanbul nuclear negotiations.

Given this tense history, what is the future outlook for the nuclear negotiations? To what extent will a compromise be possible and what actions will assuage increased confrontation and the risk of war? Answering these questions requires sound understanding of the aims, principles and strategies of Iran's nuclear program, as well as the politics of the issue, from different domestic, regional and international angles. The main discussion advanced in this chapter is that the nature of the Iranian nuclear program, including the emphasis by both sides on their own positions and also the connection with regional and international security issues, is such that it impels the two sides toward interacting with each other and within the context of striving towards a win-win (mutually beneficial) situation - meaning the relative accommodation of each other's demands. The Western side is gradually accepting the existing reality of the inevitability of uranium enrichment on Iranian soil, and in return Iran is becoming more cooperative in providing further information regarding its nuclear program.

Iran's nuclear program

The background

The Iranian nuclear program witnessed two turning points, first under the Shah's regime and second during the time of the Islamic Republic. The orientation of policies and activities related to the Iranian nuclear program indicates many similarities between Iranian nuclear policies before and after the Islamic revolution. The first attempts made by Iran to access nuclear technology go back to the late 1950s, when they had the help and encouragement of the United States.[1] The American agreement on the progress and development of nuclear infrastructures was reached when the two states were at the peak of good economic, political and military relations; Iran's strategic significance made it important for America's regional strategy. In this context, the Iranian technological and nuclear advancement was also a part of the American policy development of all dimensions of bilateral ties with Iran. Of course, contributing to this was the Eisenhower administration's "Atoms for Peace" project in 1953, along with American efforts to dominate the nuclear energy consumption market in the world.[2] Nuclear cooperation between Iran and the US continued in the 1960s, and culminated in the 1970s under President Nixon, when the expansion of political, economic and military relations prepared the ground for simultaneous growth in atomic cooperation. According to a survey carried out by the Stanford Research Institute in 1974, Iran needed 20,000 megawatts of atomic power within 20 years for its sustainable development.[3] On the same basis, the Shah's regime announced that Iran intended to implement plans for erecting atomic reactors aiming to attain

[1] Sam Roe, "An atomic threat made in America," *Chicago Tribune* (January 28, 2007) at http://www.chicagotribune.com/news/nationworld/chi-061209atoms-day1-story,0,2034260.htmlstory.

[2] Richard Kessler, "Argentina Offers Research Reactor Upgrade and New Unit for Iran," *Nucleonics Week*, Vol. 22 (January 1987): p. 4; Samini Seneviratne, "IAEA [International Atomic Energy Agency] Approves Argentine Fuel for Tehran Research Reactor," *Nuclear Fuel*, Vol. 3 (October 1988): p. 13.

[3] Svetlana Klyuchanskaya, "Background Guide: Iran's Nuclear Program" (2010), p. 5 at http://tmun.utmn.ru/wp-content/uploads/SC.pdf.

23,000 megawatts of atomic electricity during that time.[4] Subsequently, the Atomic Energy Organization of Iran (AEOI) was established in 1974 to realize this goal.

Iran's atomic cooperation with Germany and France also expanded during these years. In 1974, Germany started building two 1,200 megawatt light water nuclear reactors in Bushehr based on an agreement reached with the incumbent Iranian administration.[5] In July 1976, two other agreements were signed with the German company Kraftwerk Union (KWU) for designing, building, installing and operating two 1,300 megawatt light water power units in Bushehr and for supplying nuclear fuel of these power plants.[6] Meanwhile, another cooperation agreement was signed between Iran and Germany on the peaceful use of nuclear energy, cooperation in the field of research, technological and scientific development, education for the academic and technical staff, nuclear fuel, using nuclear energy for purposes rather than producing electricity, and so on.[7] Iran also made a cooperation agreement with France in December 1974 in the field of uranium enrichment and related financial issues, the most important part of which was the Eurodif Consortium (a joint venture for enrichment comprising France, Belgium, Spain and Italy, based in France); its implications still continue. On the basis of the agreement signed between the AEOI and the French industries, Iran and France agreed to purchase two nuclear power plants each with the capacity of 900 megawatts.[8]

The second round of Iranian attempts to develop nuclear technologies dates back to the late 1980s when Iran decided to revive its nuclear program to develop indigenous knowledge, based on a strategy for "self sufficiency" and "independence."[9] From this juncture onwards, several attempts made by Iran

[4] Kaveh L. Afrasiabi, *Iran's Nuclear Program: Debating Facts Versus Fictions* (Charleston, SC: BookSurge Publishing, 2006), p. 19.

[5] Ann Maclachlan, "Iran Seeking Way To Finish Bushehr Plant But Bonn Denies Exports," *Nucleonics Week*, Vol. 27, No. 44 (October 30, 1986): pp. 4–5.

[6] "Russia Helps Build Bushehr Nuclear Power Plant," in Foreign Broadcast Information Service (FBIS) Document NES-94–079 (April 25, 1994): p. 82.

[7] Mahdi Alikhani, "Revolution,War and Nuclear Program 1357–1377" (in Farsi), Islamic Revolution Documents Center (Farvardin 23,1385 or April 17, 2006) at http://www.irdc.ir/fa/content/5115/default.aspx.

[8] "Iran Profile: Nuclear Chronology 1957–1985," Nuclear Threat Initiative at http://www.nti.org/e_research/profiles/Iran/Nuclear/chronology_1957_1985.html.

[9] Seyed Kamal and Mohammad Rafiee, "A Study on the Global Process of Disarmament Strategy and Iran's Interaction"(in Farsi) in Seyed Hossien Mosavian (ed.), *The*

to continue its cooperation with Germany, France and some other European states, including Sweden, were stymied under US pressure.[10] Thus, Iran turned toward the East, namely, China and Russia. In 1991 primary agreements were made with China to buy a 20 megawatt reactor, followed by an agreement reached with Iran and China in December 1992 to build one or two atomic reactors with a 330 megawatt capacity.[11] However, once more under US pressure, China refused to cooperate with Iran.[12] As a result, Iran moved toward Russia and an agreement on atomic cooperation was reached on January 8, 1995 between Iran and Russia with contributions by the Russian company AtomStoryExport.[13] The agreement goes beyond cooperation toward a start-up atomic reactor in Bushehr, and establishes contracts for training Iranian engineers and technicians. Iranian-Russian nuclear cooperation witnessed many hurdles, but finally on November 30, 2010 the first Iranian atomic reactor was successfully launched at Bushehr.

A comparative study of Iranian nuclear activities reveals that the Iranian nuclear program exceeds mere access to nuclear energy and economic growth and is concerned with more substantial issues, like mastering indigenous knowledge of nuclear technology; in other words, having access to an "independent nuclear fuel cycle." This is the embodiment of progress and regional and international prestige for Iran, as the nuclear program has become a symbol of national unity, the country's desire for development and advancement, as well as resistance to foreign powers' unacceptable demands. A comparison between the Shah's Iran and Iran post-Revolution indicates there are three initiatives to be taken into consideration, the first of which was the announcement of the twenty-year plan aiming to achieve 20,000 megawatts in atomic

Additional Protocol and the Islamic Republic of Iran's Strategy, publication by the Center for Strategic Research (Tehran: Winter 2007), p. 217.

[10] R.K. Ramazani, "Making a U.S.-Iranian nuclear deal," *Agence Global* (Greensboro, NC: November 9, 2009) at http://www.agenceglobal.com/Article.asp?Id=2187; Mark Hibbs, "U.S. in 1983 Stopped IAEA from Helping Iran Make UF6," *Nuclear Fuel*, Vol. 28, No. 16 (August 4, 2003) at http://www.webcitation.org/query?url=http://www.geocities.com/thelasian/nucleafuel_iran.html&date=2009-10-25+23:00:25.

[11] Mahdi Alikhani, "China's Nuclear Exports and Assistance to Iran," Nuclear Threat Initiative at http://www.nti.org/db/china/niranpos.htm.

[12] See "Nuclear Activities of Iran after the Islamic Revolution..." (in Farsi) at http://www.irannuclear.net/node/111.

[13] *Islamic Republic News Agency (IRNA)* (Iran: December 1, 2010).

electricity, proposed by the Shah's regime for 1974–94. This goal was attained only recently as the Islamic Consultative Assembly (Iranian parliament) ratified the Iranian government plan for establishing twenty more nuclear sites.[14]

Second is the planned Iranian investment for participating in nuclear fuel production with other countries, under which, in the time of the Shah, agreements were reached with the US and some European countries. For instance, Iran signed an agreement to purchase 10 percent of the shares of Eurodif to reach independently enriched uranium fuel with a degree of 3.5 percent (Low Enriched Uranium) to supply fuel for the national nuclear program and various reactors in mind until the 1990s. The Islamic Republic's dealings are different, however, as one of Iran's basic goals in the nuclear talks since 2003 has been establishing a consortium with European countries or Russia to supply fuel, this time on Iranian soil.[15]

The final consideration is the kind of Iranian policies that insist Iranian engineers and technicians be trained through local expertise or through knowledge transfer. In the time of the Shah, agreements were signed with the French and Americans. For example, the agreement with the Massachusetts Institute of Technology (MIT) to train Iranian personnel or the one signed with Eurodif resulted in Iranian engineers in France.[16] The Islamic Republic of Iran, on the basis of an agreement with Russia, was able to have many Iranian technicians and engineers trained in Russian institutions or at the nuclear site of Bushehr.[17] In fact, at present, Iran's experts have the advantage

[14] Hamid Reza Fouladgar, "2000 Megawatts of Nuclear Power Energy in the Power Industry's Cycle" (in Farsi), *Fars News* (Ordibehesht 13, 1390 or May 3, 2011) at http://www.farsnews.com/newstext.php?nn=9002131080; "Ten New Enrichment Sites Will Be Constructed" (in Farsi), *Government Information Center* (Azar 8, 1388 or November 29, 2009) at http://www.president.ir/fa/?ArtID=19078.

[15] For Iran's nuclear proposal at that time see "Remarks of President Mahmoud Ahmadinejad's at the UN General Assembly" (in Farsi), *Government's Information Center* (Shahrivar 26, 1385 or September 17, 2006) at http://www.president.ir/fa/?ArtID=2393; see also the former head of the Atomic Energy Organization of Iran Ali Akbar Salehi, "Iran's Proposal to Establish a Nuclear Fuel Consortium with Russia" (in Farsi), *Tabnak* (Shahrivar 4, 1389 or August 26, 2011) at http://www.tabnak.com/nbody.php?id=64217.

[16] Kayhan Barzegar, "The West's Myths About Iran's Nuclear Program" (in Farsi), *Tabnak* (Azar 19, 1389 or December 27, 2010) at http://www.tabnak.ir/fa/news/138838.

[17] Naser Saghafi Ameri, "Bushehr Power Plant Sets a New Trend," Institute For Strategic Research Journals (May 1, 2011) at http://www.isrjournals.ir/en/essay/183-bushehr-power-plant-sets-a-new-trend.html.

of acquiring both Western and Eastern knowledge, and they follow a national model of their own.

Aims and principles

The Iranian nuclear program, which started in the period of the Shah, displays a diverse array of ambitions for the country's economy, technology, politics and identity. Throughout the Islamic Republic its ideological and value-based aspects were enhanced.

Regarding the *economic aspects*, the most important aspect of the Iranian nuclear program is to attain "balance in energy consumption" for sustained development. Diversification of energy resources has been always a main concern for Iranian leaders.[18] As Iran is a young and growing country with a population of approximately 75 million, fossil energy (oil and gas) consumption per capita and electricity usage in Iran stand at a level higher than the world average.[19] From this perspective, atomic energy appears to be an attractive alternative, preserving fossil resources for exports and foreign currency acquisition. Meanwhile, the availability of dependable energy resources, such as atomic energy, is essential for avoiding price fluctuations in oil and gas and also for achieving economic growth and development.[20]

Regarding the *technological aspects*, the most important technological goal sought by the Iranian nuclear program is having access to advanced atomic knowledge and becoming the master of sciences. For Iran, nuclear energy exemplifies advancement, strength and scientific progress.[21] Reinforcement of technological dimensions in the scientific, industrial, research and develop-

[18] Abbas Maleki, "Iran's Nuclear File: Recommendations for the Future," *Daedalus*, Vol. 13, No. 1 (Winter 2010) p. 106; Homeira Moshirzadeh, "Discursive Foundations of Iran's Nuclear Policy," *Security Dialogue*, Vol. 38, No. 4 (December 2007): p. 524; Kayhan Barzegar, "The Iranian Quagmire: How to Move Forward," *Bulletin of the Atomic Scientists*, Vol. 66, No. 6 (Chicago, IL: December 2010): pp. 109–114.

[19] "The Energy Superpower of 2011?", *Pipeline Magazine* (January 24, 2011) at http://www.pipelinecommunity.com/Features/the-energy-superpower-of-2011.html.

[20] Mohammad Javad Zarif, "Tackling The IRAN-U.S. Crisis: The Need For a Paradigm Shift," *Journal of International Affairs*, Vol. 60, No. 2 (Spring–Summer 2007): p. 79.

[21] Mehdi Askarieh, "A Case for Sustainable Development of Nuclear Energy and a Brief Account of Iran's Nuclear Programs," in Homa Katouzian and Hossein Shahidi (eds), *Iran in the 21st Century* (New York: Routledge, 2008), p. 189.

ment, agricultural, health and treatment fields is of immense importance. One of the greatest duties of Iranian governments has been to reconcile the demands of the middle class and the academic community with the country's level of development and advancement.[22] From this perspective, nuclear energy would strengthen expectations of technological development; moreover, it legitimizes the Iranian government's development orientation.

Table. 9.1: The economic aspects of Iran's nuclear program

Concepts	*Policies*
Sustainable development	Balance of energy
Sharp growth of population and energy consumption	Diversification of the energy sources
Power and electricity	Reliable energy sources
Maintaining fossil fuel resources	Exporting for hard currency and preserving them for future generations

Table 9.2: Technological and scientific aspects of Iran's nuclear program

Concepts	*Policies*
Advanced knowledge brings strength	Determination to acquire nuclear technology
Mother technology	Industrial, agriculture, medicine, research and development, etc
Nuclear energy	Advanced country

Then there is the *political and identity aspect*, concerning mostly national pride. Being a "nuclear state" bolsters Iranian national pride, which is already buttressed by a long history and ancient civilization. In ancient times it was one of the world's main powers, which later saw its status decline due to the reign of incompetent and dysfunctional governments. Some Iranian intellectuals believe that the weakening of Iran in the seventeenth and eighteenth centuries was the result of its lack of scientific knowledge and technologies.[23] Returning to the proud past and strengthening the pillars of national power

[22] Moshirzadeh, "Discursive Foundations of Iran's Nuclear Policy," p. 524.

[23] See Sadegh Zibakalam, "Ma Chegooneh Ma Shodim," *Rozaneh* (Fall 2003).

is a dream that has always existed in Iranian society. This nostalgia has raised Iranians' expectations of their governments to push for development and progress. For this reason, the public considers the nuclear program a national and strategic project, which could win back privileged regional and international status for Iran.

There are also the *ideological and value-based aspects*. The ideological aspect refers to Iran's cynical view of Western policies as depriving Iran of nuclear technology. The Islamic revolution was ideological by nature; so, therefore, is demanding change in the global status quo and the prevailing political, economic and social trends. Concepts like resistance, independence, justice, and Western double standards color the governing elites' views regarding the nuclear program.[24] The Iranian policy of "resistance" and attaining its legitimate rights according to NPT provisions is somehow derived from the concept of resistance to the encroachment of foreign powers, in the vein of the Islamic Revolution. Emphasis on "independence", that is mastering the "independent fuel cycle", is rooted in Iranian mistrust of the international community because of the country's general exclusion and the prevention of its independent development.[25] The emphasis on "justice" in the Iranian nuclear narrative is likewise related to the concept of justice in the Islamic Revolution—the necessity to end the existing injustice in the international political-economic trends.[26] Finally, Iran's references to the West's double standards in dealing with Japan, Brazil and Argentina's access to nuclear "capability" is linked to Iran's cynical view of the West as having recourse to double standards vis-à-vis Iran.[27]

[24] Moshirzadeh, "Discursive Foundations of Iran's Nuclear Policy," p. 522.

[25] "President Ahmadinejad: Iran Does Not Need to Get the West's Trust" (in Farsi), *Strategic Review* (Ordibehesht 14, 1389 or May 4, 2010) at http://strategicreview.org/1389/02/14; also see Foreign Minister Ali Akbar Salehi in "Historical Documents Reflect the Distrust is to Western Governments," *Mehr News Agency* (Ordibehesht 24, 1390 or May 14, 2011) at http://www.mehrnews.com/fa/newsdetail.aspx?NewsID=1310813.

[26] Mahmoud Ahmadinejad, speech at the International Conference on Disarmament and Non-Proliferation, Tehran (April 17–8, 2010) at http://www.mfa.gov.ir/cms/cms/Tehranicd/en/Files/Ahmadinejad; also see "President Ahmadinejad Raised at the NPT Review Conference: America Should Not Be a Member of the IAEA Board" (in Farsi), *Jahan News* (Ordibehesht 1389 or May 2010) at http://www.jahannews.com/prtaa6n6a49n6y1.k5k4.html.

[27] Ahmadinejad, speech at the International Conference on Disarmament and Non-Proliferation.

Table 9.3: The political and identity aspects of Iran's nuclear program

Concepts	*Policies*
Nuclear power status	State's source of power/The governments' legitimacy
National and strategic issue	State's political integration
Nuclear power capability	Potential deterrence
Strategic counterweight	Brings "political parity" to the negotiations

Table 9.4: The ideological and value-based aspects of Iran's nuclear program

Concepts	*Policies*
Resistance	Solid nuclear policy in the negotiations
Independence	Enriching uranium on Iranian soil
Justice	Nuclear energy for all
Western double standards	Seeking nuclear "capability"

Iran's nuclear policy

Iran's nuclear policy has three dimensions: cooperating with the International Atomic Energy Agency (IAEA), pursuing the issues of energy security, and following regional and global nuclear disarmament.

Cooperation with the IAEA by Iran is aimed at creating transparency and emphasizing the peaceful nature of Iran's nuclear program.[28] It also focuses on avoiding any confrontation and tension regarding the nuclear issue. Unlike Israel's nuclear strategy, which is based on "nuclear opacity,"[29] Iran's nuclear activities are fully transparent and fall within the framework of the NPT.[30]

[28] Interview with Chief of Iran's Atomic Energy Organization Ali Akbar Salehi, *CBS News* (April 13, 2010) at http://www.cbsnews.com/stories/2010/04/13/world/main6390463.shtml.

[29] Avner Cohen, *Israel and the Bomb* (New York: Columbia University Press, 1998); Avner Cohen, "Israel's Nuclear Opacity: a Political Genealogy," *The Dynamics of Middle East Nuclear Proliferation*, Vol. 66 (2001) at http://members.tripod.com/alabasters_archive/nuclear_opacity.html#contents.

[30] See "Visit Nuclear Facilities, Sign of Iran's Nuclear Program Transparency" (in Farsi), *Dari IRIB* (Dey 25, 1389 or January 15, 2011) at http://dari.irib.ir/... /29295-2011-01-16-13-04-34.html; Foreign Minister Ali Akbar Salehi, "Iran's

While Iran's cooperation with the IAEA is an opportunity, it brings about some challenges that force Iran to cooperate with the IAEA only in the context of its safeguards agreement.[31] Meanwhile, Iran's policy is based on the active presence of the NPT and further strengthening the balance of its three main principles: non-proliferation, peaceful use of nuclear energy, and comprehensive disarmament.[32] From this perspective, the NPT rules and regulations are such that they guarantee the legitimacy and peaceful nature of Iran's nuclear program. Meanwhile, compliance with NPT regulations decreases political pressures on Iran, which are mainly the consequence of existing imbalances in these principles. According to this argument, the West's excessive stress upon the principle of "non-proliferation" overshadows the other two principles. A more balanced assessment of all three elements could generate an appropriate atmosphere for reducing political pressures on Iran. According to this strategy, it is unlikely that Iran will withdraw from the NPT, although some on the inside are skeptical about the NPT's efficiency and argue that Iran's membership will only exert pressure on Iran, and so Iran should withdraw from the Treaty. The dominant policy view, however, is that withdrawal from the NPT would undermine legal, technical, and peaceful aspects of Iran's nuclear program and embolden political pressures by permitting its international critics to focus on potential military dimensions of the program.

In pursuit of *energy security*, Iran presses for comprehensive access to peaceful nuclear energy and justice in the consumption of nuclear energy at the regional and international level. The main point of disagreement between Iran and the West emerged as a result of two conflicting mindsets: "peaceful" nuclear energy, and atomic energy for "weaponry" and providing "deterrence." Iran believes that the US-dominated West intends to monopolistically control nuclear energy, using non-proliferation only as a pretext, thereby ignoring just principles for nuclear energy in attaining sustainable development.[33] If this is indeed the case, then it might be necessary to further reconcile the NPT prin-

---

Nuclear Facilities Are Open to Chinese Experts" (in Farsi), *IRNA* (Khordad 2, 1390 or May 23, 2011) at http://www.irna.ir/NewsShow.aspx?NID=30399294.

[31] Mohammad Javad Zarif, "Tackling The IRAN-U.S. Crisis": p. 75.

[32] Kayhan Barzegar, "The Balance of Power in the Persian Gulf: An Iranian View", *Middle East Policy*, Vol. 17, No. 3 (Fall 2010): p. 83.

[33] Ahmadinejad, speech at NPT Review Conference, New York (May 3–28, 2010) at http://english.irib.ir/inded.php?option=com_k2&view=item&id=6048:text-of-president-ahmadinejads-address-to-the-un-npt-preview-conference&Itemid=182.

Table 9.5: Iran's nuclear policy and the West's position

Iran's Nuclear Policy	*The West's (US's) Position*
Emphasizing the "peaceful energy" and "disarmament" themes	Emphasizing the "weaponization," "deterrence" and "proliferation" themes
Regional peace and stability approach	Regional threat approach
All-out nuclear disarmament in the Middle East	Excluding Israel's nulear arsenals (seeking nuclear disarmament in the Persian Gulf)
Connoting the Middle East's disarmament to global disarmament	Connecting the Middle East's security to global security (nuclear terrorism)
Balancing energy security	Energy security for the US and the West
Nuclear democracy, general use and transformation of nuclear technology	Monopolization of nuclear fuel and limitation of transferring nuclear technology

ciples of disarmament, proliferation and peaceful cooperation.[34] To this end, Iran demands the expansion of peaceful nuclear activities in the region, at several junctures supporting nuclear programs in other Persian Gulf states.[35] It has even declared readiness to share expertise, export nuclear technology and cooperate closely in the nuclear field.[36] But on the contrary, the Western and American standpoint principally emphasizes the threat posed by the Iranian nuclear program, implying a strategy of "deterrence."[37] A powerful and nuclear Iran would disturb the prevalent regional political-security order based on a "balance of power." The fear is that Iran would threaten existing

[34] Akbar Salehi, speech at the International Conference on Disarmament and Non-Proliferation, Tehran (April 17–8, 2010) at http://www.mfa.gov.ir/.

[35] Ali Larijani, "Iran's Strategy is to Expand Relations with Regional Countries" (in Farsi), *Ebtekar News* (Aban 2, 1387 or October 23, 2008) at http://www.ebtekarnews.com/ebtekar/News.aspx?NID=39060.

[36] Ramin Mehmanparast, "Iran is Ready to Cooperate with Regional Countries" (in Farsi), *IRNA* (Bahman 19, 1389 or February 9, 2011) at http://www2.irna.com/ar/news/view/line-7/8911182090112417.htm.

[37] Jeffrey Goldberg, "The Point of No Return," *The Atlantic* (Washington, DC: September 2010); Farid Zakaria, "Containing a Nuclear Iran," *Newsweek* (October 3, 2009) at http://www.newsweek.com/2009/10/03/containing-a-nuclear-iran.html; Kayhan Barzegar, "Win-Win Strategy in Iran's Nuclear Program," *Iranian Diplomacy* (January 1, 2011).

energy security arrangements by imposing its will on oil prices and exerting pressure on conservative Arab governments allied with the US.[38] This possibility challenges the American aim to keep Arab allies strong by placing an atomic umbrella over the Persian Gulf.[39]

A strategy for *regional and global disarmament* acquires legitimacy for Iran's nuclear program at the level of global public opinion, and at the same time intensifies political pressure on the main adversary Israel, which is not a signatory to the NPT. This policy has two dimensions: first, elimination of the "deterrence" and "weaponization" narrative, and second, a comprehensive disarmament of the Middle East. Iran has already excluded the concept of "nuclear deterrence" from its defense strategy, focusing only on conventional military deterrence.[40] Iran believes that since regional security in the Middle East is so interconnected, the country can easily counter military threats posed by the United States and Israel by resorting to conventional and asymmetric warfare.[41] Military exercises and enhancement of missile systems are part of this defense strategy. President Ahmadinejad reiterated this policy during the Tehran Nuclear Disarmament Conference (April 17–18, 2010),[42] and Supreme Leader Ayatollah Seyyed Ali Khamenei, in a letter addressed to the conference, declared the use of nuclear weapons *haram* (religiously banned).[43]

As for comprehensive disarmament in the Middle East, Iran initiated the Middle East Disarmament Plan in 1974,[44] and one of the main components

[38] See Kayhan Barzegar, "Stephen Walt on the U.S., Iran, and the New Balance of Power in the Persian Gulf," paper at the Belfer Center for Science and International Affairs (Cambridge, MA: August 5, 2008).

[39] "US warns Iran on nuclear pursuit," *Al Jazeera* (February 15, 2010) at http://english.aljazeera.net/news/middleeast/2010/02/2010214172432220260.html.

[40] Mahmoud Ahmadinejad "Today, Iran is a Nuclear State, We Reached the Point of No Return" (in Farsi), *IRNA* (Farvardin 20, 1389 or April 9, 2010).

[41] "Sepah (the IRGC) will Respond to the Enemy in the High Seas"(in Farsi), *Fars* News (Ordibehesht 2, 1390 or April 22, 2011) at http://www.farsnews.com/newstext.php?nn=9001312352.

[42] Ahmadinejad, speech at the International Conference on Disarmament and Non-Proliferation.

[43] Ayatollah Ali Khamenei, speech at the International Conference on Nuclear Disarmament, Tehran (April 17, 2010) at http://www.mfa.gov.ir/cms/cms/Tehranicd/en/Files/SupremeLeader.

[44] David Sanger, "Iran agrees to inspection of military base," *New York Times* (January 6, 2005) at http://www.nytimes.com/2005/01/06/politics/06nuke.html?oref=login&pagewanted=all.

of Iran's nuclear strategy at present is to pursue comprehensive disarmament. Such a strategy has received the active support of regional actors such as Egypt, Turkey, and Saudi Arabia, enhancing Iran's nuclear position and challenging Israel's nuclear program by pressuring Tel Aviv to sign the NPT. Iran believes that Israel poses the most imminent security threat to the region and is the origin of a conventional military arms race.[45] Iran's strategy in this regard is to form a new bloc supporting comprehensive disarmament in the region.[46] In this respect, Iran held its first disarmament conference on April 17–18, 2010 with the motto of "nuclear energy for all, nuclear weapon for none." Accordingly, President Ahmadinejad described regional and global comprehensive disarmament as the basis of Iran's nuclear policy in the NPT review conference on May 3, 2010. The follow-up disarmament conference was held on June 12–13, 2011 in Tehran.

Iran's nuclear program and domestic politics

At the domestic level, there is something of a consensus among the ruling elites on two issues. Foremost, Iran's nuclear program must not be used for military objectives. Additionally, Iran should acquire nuclear energy and power "capability." Here, acquiring capability does not necessarily mean access to a nuclear arsenal. Maintaining such "capacity" instead means the international community's acceptance of Iran as a nuclear state without weaponization.

"Preserving the independent fuel cycle" on Iranian soil gradually became the consensus point for all political tendencies in the domestic decision-making process for Iran's nuclear activities. The Iranian nuclear program strengthens the legitimacy of the political system and, consequently, the pillars of power among the governing political-security elites. Therefore, all political groups, however different their approaches may be, unanimously promote it.[47]

[45] Alaeddin Boroujerdi, head of the Majlis Commission on Foreign Policy and National Security, "The entrance of Israeli submarines in the Persian Gulf is a threat" (in Farsi), *Hamshahri Online* (Khordad 10, 1389 or May 31, 2010) at: http://www.hamshahrionline.ir/news-108669.aspx.

[46] "The Formation of the Resistance Bloc is the Result of the Great Awakening in the Middle East" (in Farsi) *Mardomsalari* (Farvardin 23, 1390 or April 12, 2011); Kayhan Barzegar, "Turkey-Iran Relations: Implications for Israel's Regional Strategies" (in Farsi), *Islamic World Strategic Studies*, Vol. 11, No. 44 (Winter 1389).

[47] Hossein Mousavian, "Iran's Nuclear Crisis and the Way Out: Part 1," *Iran Review* (February 14, 2011) at http://www.iranreview.org/content/Documents/Iran%E2%80%99s_Nuclear_Crisis_and_the_Way_Out_Part_1.htm.

The importance of the nuclear issue is rooted in the need for legitimacy for the "Iranian state." In Iranian political culture, critics have always blamed governments for failure in advancing development. In contemporary history, the Iranian nuclear program is the sole issue which is unanimously considered by Iranian society as the starting point for technological advance, sustainable economic development, national pride and cohesiveness, strengthened regional and international prestige, and military deterrence. Therefore, all domestic political groups, regardless of political inclination, appear to be engaged in the Iranian nuclear program. For instance, development-oriented groups focus attention on sustainable development, economic growth and technological advancement.[48] Nationalist groups consider the issue of increased national power, regional and international prestige and deterrence,[49] and ideological and value-oriented groups espouse values like independence, resistance and identity.[50] In fact, the nuclear program is the common denominator of all national demands of the Iranians.

At the domestic level, the Iranian nuclear program has led to a homogenization of the political spectrum. There is consensus among elites over such things as interaction between internal power institutions, interaction with the people and public opinion, the significance of regional issues, and finally, engagement in direct talks with the US over the nuclear program.[51]

First, the Iranian nuclear program has led to an interaction between power institutions in domestic politics on the one hand and public opinion and the people on the other. This more than anything else has markedly strengthened the working of Iranian democracy, in addition to keeping a balance among nuclear decision making, domestic power institutions such as the Supreme

[48] Hasan Rohani *et al.*, "Iran's National Security and Economic System," publication by the Center for Strategic Research (Tehran: 2010); Mohammad Nahavandian, "We Should Negotiate for Lifting the Sanctions" (in Farsi), *Strategic Review* (Azar 10, 1389 or December 10, 2010) at http://strategicreview.org/1389/09/10/

[49] Davood Hermidas Bavand, "Iran and the Global Nuclear System" (in Farsi), *Aftab News* (Bahman 27, 1384 or February 16, 2006) at http://www.aftabnews.ir/vdcg3w9t.ak9qw4prra.html; "Iran's Nuclear Case: Rise and Fall" (in Farsi), *Islamic Revolution Documents Center* (Farvardin 28, 1387 or April 18, 2008) at http://www.irdc.ir/fa/content/6266/default.aspx.

[50] See the works of Mehdi Mohammadi at his blog http://mohammadikayhan.blogfa.com/cat-1.aspx; Mehdi Mohammadi, "A Few Points About a Day of Negotiation," *Kayhan* (2 Bahman 1389 or January 22, 2011).

[51] See Kayhan Barzegar, "Iran's Nuclear Program and Iran-US Strategic Dialogue" (in Farsi), *Security Afaq Quarterly*, Vol. 3, No. 6 (Spring 2010).

Leader's Office, the Presidential Office, the Supreme National Security Council, the Foreign Ministry, the AEOI, the Islamic Consultative Assembly, and security and intelligence services. Each of these institutions, according to its legal position and political importance in Iran's power structure, plays a role in the nuclear program's implementation. In accordance with the Iranian constitution, general and strategic policies are determined by the Supreme Leader and communicated to the government and Presidential Office for implementation. The Islamic Consultative Assembly has to endorse contracts and carries Iran's international responsibility. Security and intelligence institutions are charged with physically protecting nuclear facilities and the safekeeping of the Iranian nuclear intelligence. The Supreme National Security Council is tasked with keeping a balance in the process, whereby all the aforementioned institutions are represented. The Iranian President is the chief of the National Security Council.

Another aspect of balance within the Iranian decision-making process is the supportive role of the Islamic Consultative Assembly. The Iranian government's acceptance of a fuel exchange in Turkey as part of the Geneva talks in October 2009 was opposed by the major current within the assembly, eventually halting further negotiations on the issue.[52] The main barrier to this agreement was the lack of a strategic role dedicated to the assembly in affairs related to Iranian national interests and security. The ousting of former Foreign Minister Manuchehr Mottaki in December 2010 aimed at homogenizing nuclear decision making in the Ahmadinejad administration and also made clear Iranian institutions' demands for negotiating parity and so a balance in the nuclear talks process.[53] Eventually, with the appointment of the new Foreign Minister Ali Akbar Salehi, the government objective of attaining a political consensus over the nuclear program was achieved and ushered in the assembly's vote of confidence.[54] As a result, the nuclear program has led to

[52] Ali Larijani, "Tough Opposition in the Majlis for Sending the Uranium Out" (in Farsi) (Aban 4, 1388 or November 25, 2009); Robert Dreyfuss, "Iran bombshell: US reveals secret facility," *The Nation* (New York: September 25, 2009); Robert Dreyfuss, "Iran sanctions: not just useless but counterproductive," *The Nation* (New York: June 9, 2010).

[53] "Parliament's Reaction to Dismissing Manouchehr Mottaki" (in Farsi), *Khabar Online* (Azar 28, 1389 or December 19, 2010) at http://www.khabaronline.ir/news-116992.aspx.

[54] "Parliament's Vote of Confidence to Ali Akbar Salehi" (in Farsi), *IRNA* (Bahman 10, 1389 or January 30, 2011) at http://www.irna.ir/Display.aspx?NID=030213515.

Table 9.6: Iran's decision-making process

Decision-making process (key concepts)	1. Various balancing constituencies. 2. A multipolar decision-making process. 3. Complicated and slow, but safe, wise and strategic.
Main actors	1. Supreme National Security Council: key decision maker. 2. Foreign Ministry: legal and political issues. 3. Atomic Energy Organization of Iran (AEOI): technical and logistical issues. 4. Iranian Revolutionary Guards Commission (IRGC): protecting facilities and tackling foreign military threats. 5. Intelligence Ministry: protecting the transference of intelligence and nuclear scientists.
Other actors	1. The Islamic Consultative Assembly (the Iranian Parliament): equating other executive powers and public. 2. The Guardian Council (ratifying the Additional Protocol (meant to enhance IAEA monitoring); any consortium plans). 3. The Expediency Council (ratifying any of the consortium plans; the Additional Protocol; other Strategic Plans). 4. The media: balancing public opinion.

more interaction, enhanced roles and increased interdependence of varying power institutions in Iran's domestic politics.

Meanwhile, the Iranian nuclear program has led to the convergence of public and governmental views on an important issue of Iran's foreign policy and international relations. The latest polls show that a majority supports a peaceful nuclear program.[55] This itself makes the Iranian nuclear issue legitimate in regional and even global public opinion. Societal support not only ensures the continuation of Iran's peaceful nuclear program, but also indicates the extent to which the Iranian people are sensitive to defending legitimate interests and

[55] An opinion poll conducted by the US RAND Corporation showed that 87 percent of Iranians strongly favor Iranian development of nuclear energy for civilian use. *Tehran Times* (May 3, 2011) at http://www.tehrantimes.com/index_View.asp?code=239876.

policies based on bolstering national power. This has brought about political homogeneity in the domestic decision-making process.

Another issue on the domestic political scene has been the way political-security elites view regional issues within the framework of the Iranian nuclear program. Primarily, the Iranian nuclear program has led to a paradigm shift, and also a shift in its priorities in the conduct of foreign policy in the minds of the decision-making elites. Previously, priority was given to regional and ideological issues, including Hezbollah and Palestine. However, with the introduction of the nuclear issue into Iranian foreign policy, more emphasis was put on the strategic interests and issues of national security and power in regulating regional ties.[56] As will be elaborated in the next section, the active presence of the nuclear program in Iranian politics has the potential to employ the most important means of Iran's national power, the geopolitical element (regionalization of Iran's nuclear program)[57] and the ideological element (strategizing ideologies).[58]

Table 9.7: Iran's nuclear program: discourses, concepts and principles

Discourses	*Discussion*	*Concepts/principles*
A. Developmental approach (the mainstream discourse)	Modernity, progress	Independence, resistance, justice, double standards, inalienable rights
B. Nuclear "capability"	Popular/consensus	Regional power status
C. Weaponization	Unpopular	Against humanity, *haram* (religiously banned)

Meanwhile, reviewing Iranian defense and security strategies reveals that the Iranian nuclear program has introduced and strengthened the concept of

56 Ismail Kosari (Member of National Security and Foreign Policy Commission of the Parliament, "Iran's Successful Performance in Foreign Policy" (in Farsi) (Farvardin 6, 1388 or March 26, 2009) at http://www.irdiplomacy.ir/Modules/News/Phtml/News.PrintVersion.Html.php?Lang=fa&TypeId=40&NewsId=4237. Also, Mohammad Sadegh Kharrazi (Iran's former ambassador to France 2002–6)," Nuclear incident" (in Farsi), *Personal Website* (Tir 23, 1387 or July 14,2008) at http://kharazi.ir/fa/news/6/bodyView/17/0/%D.html.

57 Mehdi Mohammadi, "Read the text of the modality," *Vatan Emrooz* (Tehran: Farvardin 17, 1389 or April 2, 2010).

58 Kayhan Barzegar, "Iran's Foreign Policy After Saddam," *The Washington Quarterly* (January 2010): pp. 178–80.

"interconnected security" in the Iranian power structure. On the same basis, Iran has connected the security of its nuclear power plant to the security of the region and has warned states both in and out of the region that any contribution to and cooperation with invading foreign forces against its nuclear facilities would be met with an offensive Iranian reaction.[59] In other words, insecurity for Iran is equivalent to insecurity for the region, and so, for the world. With this defense policy, Iran has defined its strategy as "offensive defense," the basis of which is active military engagement and confrontation with any disturbances in regional trade, goods transport and energy export. Iran has made clear that it would employ all possible means at its disposal, including the use of its political-security potentials and friendly political groups in the region.[60]

In this context, the Iranian nuclear program has changed the views and perspectives of the Iranian political-security elites, as well as the Iranian public, on relations with conservative Arab states especially in the Persian Gulf. During the past few years, the Arab countries of the region have used a double standard when dealing with Iran's nuclear program. On one hand, they accept the legitimacy of Iranian access to peaceful nuclear energy since they themselves are currently pursuing this technology for sustainable development and prestige;[61] Arab states in the region have signed several nuclear agreements with Western countries—the UAE with France and South Korea, Saudi Arabia with the US, Kuwait with Japan, Qatar with South Korea, and so on—and the first nuclear reactor of the region is due to operate in Abu Dhabi with the help of France in 2017.[62] On the other, the conservative Arab regimes support robust sanctions and, in some cases, military action against Iran, as several Wikileaks documents have shown.[63] The conservative Arab regimes have dis-

[59] Ayatollah Ali Khamenei said: "If the U.S. unwisely follows through on its military threats against Iran, the repercussions of that attack will not be restricted to the region." *Mehr News* (August 19, 2010).

[60] Kayhan Barzegar, "Iran's Foreign Policy Strategy after Saddam", *The Washington Quarterly*, Vol. 33, No. 1 (2010): p. 180.

[61] Kayhan Barzegar, "Atoms for Sustainable Development and Prestige," strategic report, Scientific Research and Middle East Strategic Studies Center (June 2010).

[62] Matthew Fuhrman, "Oil for nukes: mostly a bad idea," *Christian Science Monitor* (Boston, MA: February 29, 2008) at http://www.csmonitor.com/2008/0229/p09s02-coop.html.

[63] "Leaks of American Diplomatic Messages and the Regional Concerns Toward Iran" (in Farsi), *BBC Persian* (November 29, 2010) at http://www.bbc.co.uk/persian/world/2010/11/101128u03wikileaksiran.shtml.

played a clear role in supporting the UN Security Council sanctions, whether practically through limiting financial cooperation and trade with Iran, or theoretically through political rhetoric threatening to inspect Iranian merchant ships as part of Security Council sanctions.[64] The Arab regimes' dual stance vis-à-vis the Iranian nuclear program has emboldened cynical sentiments among internal political groups toward the conservative Arab regimes in the region, resulting in the convergence of the views of political-security elites toward these Arab regimes.

The Iranian nuclear program has led to a relative political consensus among the political-security elites in Iran on the necessity of entering strategic talks with the US.[65] Negotiation with the US has always been a critical issue in Iran's domestic politics, having influenced the course of interaction between different political groups. Beyond all existing ideological conflicts and past bitter hostilities, the Iranian nuclear program in the eyes of Iran's political-security elites is the sole issue that has the potential for providing a basis for the promotion of direct talks between Iran and the US. Iran and the US have independent and often conflicting views on bilateral issues including the handling of the regional crises in Afghanistan, Iraq, Palestine and Lebanon, combating terrorism, conducting the Middle East peace process, human rights issues, energy security, and so on. For example, Iran considers the presence of American troops in Afghanistan and Iraq as one of the main causes for the spread of violent activities in these regions; this is very different from the American point of view, which holds that there is a military solution to solving the crisis. Also, Iran's view is that what brings security for the US in Iraq has an adverse security effect on Iran. Furthermore Iran recognizes Hamas and Hezbollah as political movements and resistance groups while the US calls them terrorists.[66]

[64] See Ariel Farrar-Wellman and Robert Frasco, "United Arab Emirates-Iran Foreign Relations," *Iran Tracker*, The American Enterprise Institute (July 21, 2010) at http://www.irantracker.org/foreign-relations/united-arab-emirates-iran-foreign-relations; "Clinton seeks to shore up support from Arab allies against Iran," *Fox News* (New York: January 9, 2011) at http://www.foxnews.com/politics/2011/01/09/clinton-seeks-shore-support-arab-allies-iran/#ixzz1NiovVPpP.

[65] Kayhan Barzegar, "The Nuclear Program and Iran-US Mutual Strategic Need," strategic report, Middle East Strategic Studies Center (October 2010) at http://fa.merc.ir/archive/article/tabId/62/articleType/ArticleView/articleId/1071/—.aspx.

[66] See Kayhan Barzegar, "Iran, New Iraq, and the Persian Gulf Political-Security System," publication by the Center for Strategic Research (Tehran: Fall 2008), pp. 158–62.

Yet, one can argue that the nature of Iran's nuclear program is such that it brings Iran and the US closer to each other and increases the possibility of direct talks between the two sides. From a national and strategic point of view, the Iranian nuclear program is the only issue on which there is an internal consensus among the elites of both sides for direct bilateral talks. In the case of Iran, as discussed above, the political consensus among the political-security elites for preserving an independent fuel cycle and achieving a win-win outcome that protects Iran's national interests and security convinces the elites that direct talks could potentially settle the Iran-West nuclear standoff. In the case of the US, imagining the possibility of the Iranian nuclear program diverted for a military purpose, and the exigency to manage and control the Iranian nuclear program, encourage the Americans toward direct negotiations with Iran. In other words, the emphasis placed by both sides on their own interests leads them to choose between resolving the issue diplomatically and engaging in conflict. As the evidence shows, both sides favor interaction since war comes with its own complications, especially given the current political and democratic movements across the Arab world. Both sides have adopted policies in such a way that they are unable to negate the other side's demands entirely. Iran could not relinquish its claim to the independent fuel cycle as it has already endured considerable political and economic costs. The US, also, could not tolerate a nuclear-armed Iran. The balancing point therefore is to accept each other's basic demands in a relative manner toward a win-win situation. Such a condition for the first time has bound Iranian and American interests together—the result of consensus among the respective political elites to engage in direct talks.

Iran's nuclear program and regional politics

Another feature of the Iranian nuclear program is its interdependence with regional politics. First, the nature of Iran's nuclear program is linked to the political, economic and security issues of the region, and also to the interests of regional and trans-regional players. Second, Iran strives to regionalize its nuclear program.

Regarding the regional effects of Iran's nuclear program, four parameters are significant:

Energy security

Iran believes that access to peaceful nuclear energy is necessary for the sustainable development of the region, and regional states including the Arab coun-

Table 9.8: Politics of Iran's nuclear program: Hardliners/Conservatives/Reformists

	Hardliners (ideological/pragmatic)	*Conservatives (pragmatic)*	*Reformists (moderate)*
Sources of power and legitimacy	– The President – The Islamic Consultative Assembly (ICA) – IRGC – Political–security bodies (militarized) – The Judiciary – The Guardian Council – The Expert ICA – Friday prayers – The media – The masses	– The Expediency Council – Economic bodies – Oil and gas sections – Industries – Executive intellectuals – Universities – The middle class	– The media circle – Universities – Academic intellectuals – Students – The middle class
Iran's place in the world	– Already a regional power and a global actor	– A rising regional political-economic power	– A rising political– economic power
How Iran sees the world/ foreign policy	– East-looking policy – The multipolar world, the Non-Aligned Movement (NAM) countries	– The West/East multipolar world – The Islamic world	– The West multipolar world

Sanctions and UNSC resolutions	– is an opportunity	– is a threat/an opportunity	– is a Threat
Regional policy	– Proactive policy	– Detente/confidence– building	– Detente
Suspending the enrichment of uranium on Iranian soil	– No suspension – Proactive diplomacy	– No suspension (only voluntary and limited suspension for confidence– building) – Accommodating diplomacy	– No Suspension – Confidence– building diplomacy
Acquiring nuclear energy "capability"	– Positive	– Positive	– Positive
Weaponization	– Negative	– Negative	– Negative

tries of the Persian Gulf, Turkey, Egypt, etc., have accepted this logic and are proceeding accordingly. The Iranian model for making peaceful use of nuclear energy and achieving sustainable development, technological progress and prestige in the region have caught the attention of the Arab states, especially the Persian Gulf countries. The issue of acquiring nuclear energy for purposes of sustainable development plays an important role in the future advancement of regional Arab states. Supposing that the supply of fossil fuels will come to an end in the coming decades, the future would be dark and unreliable in terms of economic growth and development, and as a consequence dynamism, financial and commercial centrality, foreign investment and tourism would be lost. Given such a projection, regional states have started to think about possible substitutes, including nuclear energy.

On the other hand, access to advanced technologies, including nuclear power, could be a source of prestige and a basis for strengthening national identity and solidarity for the Arab states of the Persian Gulf. In past years, there has been serious competition between these states over access to sophisticated technologies, with their operating high profile plans to gain the attention of international public opinion. From this angle, taking action to operate the Bushehr power plant has recharged regional ambitions to use nuclear energy for sustainable development, technological advancement and prestige. Following the Bushehr plant, several agreements were signed between the regional states and foreign countries, one of which is the Kuwait-Japan agreement for building an atomic power plant, mentioned below.[67] The Iranian model for peaceful nuclear energy has advanced the issue of energy security in the political-economic strategies of different states in the region.

Arms race

The Iran nuclear issue is interconnected with the arms race in the region. The main narrative, which gained credence in the West, in Israel, and among conservative Arab regimes, is that a nuclear Iran will have a domino effect encouraging the nuclearization of Saudi Arabia, Turkey and Egypt, leading to an arms race.[68] Distorting the Iranian nuclear program as a threat to regional

[67] "Japan, Kuwait agree on nuclear energy cooperation," *Gulf News* (Dubai: September 9, 2010) at http://gulfnews.com/business/general/japan-kuwait-agree-on-nuclear-energycooperation-1.679474? localLinksEnabled=false&utm_source=Feeds&localLinksEnabled=false&utm_source=Feeds&utm_medium=RSS&utm_term.

[68] Judith S. Yaphe and Charles D. Lutes, "Reassessing the Implications of a Nuclear-

security, the West's pointed plan focuses mainly on the theme of "weaponization" of Iran's nuclear program, even though the Iranian nuclear program, as discussed above, is based on "peaceful" nuclear energy and "deterrence." There are of course counter arguments in the West that even weaponization of Iran's nuclear program does not necessarily amount to proliferation of nuclear weapons in the region.[69] For example, Israel's weapons program in the 1960s and 1970s did not necessarily lead Saudi Arabia or Egypt to acquire nuclear weapons, although there were serious hostilities and several wars between Israel and the Arabs at the time.

Nuclear terrorism

As terrorism emerged as a major concept in the US national security strategy,[70] it somehow became linked with nuclear activities in the Middle East. For the Obama administration "the greatest threat against the US national security in the short, medium or long term, is the possibility of access to nuclear materials or weapons by violent groups (mainly in the Middle East)."[71] Holding the Nuclear Security Summit in Washington was a turning point in this regard. The link between the issues of world terrorism and nuclear terrorism in American national security, as well as global security, is related to the political-security issues in the Middle East and Iran's nuclear program.[72] Here, the US argues that Iranian access to nuclear arms could increase the danger of transferring nuclear weapons to the so-called terrorist groups such as Hamas and Hezbollah, which already enjoy close relations with Iran.[73] On the other

---

Armed Iran," Institute for National Strategic Studies, National Defense University (Washington, DC: 2005) at http://www.ndu.edu/inss/docUploaded/McNair69.pdf.

[69] Stephen M. Walt, "Iran, Arms Races, and War," *Foreign Policy* (Washington, DC: October 1, 2009).

[70] See Graham Allison, "Assessing Obama's nuclear summit," *The Daily Beast* at www.thedailybeast.com (April 17, 2010); Matthew Bunn, "Nuclear smuggling: the expert view," *The Guardian* (London: November 7, 2010).

[71] Barack Obama, speech at the Nuclear Security Summit, US Department of State, Washington, DC (April 12–13, 2010) at http://fpc.state.gov/c35775.htm.

[72] Kayhan Barzegar, "Nuclear Terrorism, Iran, and the NPR," *World Bulletin* (April 26, 2010) at http://www.worldbulletin.net/news_detail.php?id=57644.

[73] Eric S. Edelman, Andrew F. Krepinevich Jr. and Evan Braden Montgomery, "The Dangers of a Nuclear Iran," *Foreign Policy* (Washington, DC: January–February

hand, the West cooperates with the Arab states in the region on the transfer of nuclear technologies and materials for reactors in construction.[74] Undoubtedly, the main center of activity for radical and violent groups is in the Middle East. Therefore, the existence of nuclear materials in Arab countries always implies the danger of these groups' access to nuclear materials, which is in turn a risk for regional and world security. In other words, through, as it were, the intermediary of "nuclear terrorism" the US has linked its national security to the security of the Middle East.

Regional balance of power

The challenge seen by the West in Iran's nuclear program relates very much to the issue of preserving the balance of power in the Middle East. From the West's perspective, especially that of the United States, a nuclear-armed Iran will disturb the traditional balance of power to the detriment of the West and its traditional regional allies.[75] Some American strategists maintain that in order to keep its hegemonic control over existing security-military currents in the Middle East, the US has to gain control over Iran's nuclear program.[76] The idea of an "existential threat" to Israeli security from Iran's nuclear program[77] is also rooted in the so-called "balance of power", as keeping the Israeli nuclear monopoly over the region somehow relates to the Iranian nuclear program.[78] From this perspective, if Iran acquires nuclear weapons, it would challenge the existence of the Israeli regime and bring the whole region face-to-face with a crisis of political-security and legitimacy.[79] The US would then see itself com-

---

2011); Massimo Calabresi, "Iran's nuclear threat," *Time* (New York: March 8, 2003) at http://www.time.com/time/world/article/0,8599,430649,00.html.

[74] Barzegar, "Atoms for Sustainable Development and Prestige."

[75] See "Nicholas Burns, Former Under Secretary of State for Political Affairs, Speaks on the Future of the U.S.–Iran Relationship," press release, Harvard University and Belfer Center for Science and International Affairs (Cambridge, MA: March 11, 2008).

[76] Noam Chomsky, "The Iranian Threat" ZCommunications (June 29, 2010) at http://www.zcommunications. org/the-iranian-threat-by-noam-chomsky.

[77] See Chuck Freilich, "The United States, Israel, and Iran: Defusing an Existential Threat," *Arms Control Today* (Washington, DC: November 2008).

[78] See Martin Malin, "Four Reasons the US Could Get Israel to Talk About a Middle East Free of Weapons of Mass Destruction," *Christian Science Monitor* (Boston, MA: June 8, 2010).

[79] Avner Cohen, "Israel Ponders a Nuclear Iran," *Bulletin of the Atomic Scientists* (Chi-

pelled to take action and provide its allies with the necessary security guarantees to protect them against the Iranian threat. Remarks made by Secretary of State Hillary Clinton regarding the US tendency to provide a nuclear security umbrella for the regional states are to be evaluated in the same vein.[80]

The second issue is the "regionalization" of Iran's nuclear program. This Iranian strategy is founded upon the following pillars:

First, Iran's key role in connecting the security of the Middle East to international security. Since the events of September 11, the security system of the Middle East has been tied to international security by new variables like Al Qaeda terrorism, regional crises in Iraq, Afghanistan, Lebanon, Palestine and Yemen, and recently the democratic movements in the Arab world like those in Tunisia, Egypt, Libya, Bahrain, and others. Today, two major topics of international security fall within the realm of the Middle East: combating terrorism and preventing their access to WMDs, and deterring the spread of regional crises.[81] Iran is the only regional player that, ideologically and geopolitically, leaves a serious impact on the settlement of the regional issues, as none of the above goals can be attained without Iran's constructive role. From this angle, the existing situation in the Middle East presents a precious opportunity for Iran to enhance its strategic position through linking its nuclear issue to other regional concerns.[82]

Second, creating "political strategic parity" in the nuclear talks. Another reason why the Iranian nuclear program has been regionalized is to create a "political parity" in the nuclear negotiations with the great powers. By an active policy and effective presence in regional issues such as the crises in Afghanistan, Iraq, Lebanon and Palestine, Iran has tried to establish an equally vital, if not indispensible, "political role" in the nuclear talks, especially with respect to the US. Alongside this, Iran has tried to enhance its strategic value and the ability to act at the regional level in three ways:

---

cago, IL: August 17, 2010); Barzegar, "Turkey-Iran Relations: Implications for the Israel's Strategies."

80 Mark Landler and David E. Sanger, "Clinton speaks of shielding Mideast from Iran," *New York Times* (July 22, 2009) at http://www.nytimes.com/2009/07/23/world/asia/23diplo.html; Mark Landler, "Clinton raises U.S. concerns of military power in Iran," *New York Times* (February 15, 2010) at http://www.nytimes.com/2010/02/16/world/.../16diplo.html.

81 Kayhan Barzegar, "Iran, the Middle East, and International Security," *Ortadogu Etütleri*, Vol. 1, No. 1 (July 2009): pp. 28–30.

82 Mohammadi, "Read the text of the modality."

1) Strengthening the policy of alliance and coalition with friendly governments and political factions such as Syria, Hezbollah, Hamas and Shi'ite groups in Iraq. In this respect, Iran has increased its strength and influence in the region through supporting Syria and Hezbollah vis-à-vis the Rafiq Hariri tribunal, through President Ahmadinejad's visit to Lebanon, as well as through strengthening Maliki's Shi'ite government in Iraq.
2) Bolstering and supporting the activities of new actors like Turkey in regional issues and granting a vital role to these actors in the nuclear talks and regional crises. Iran's insistence on holding the nuclear talks in Turkey in January 2011 is an example of this. During his recent visit to Turkey (May 2011), President Ahmadinejad once again stressed that Istanbul is a good place for future nuclear talks.[83] With the same aim, Iran also insists on its commitment to exchange nuclear fuel on Turkish soil on the basis of the Tehran Nuclear Declaration.
3) Finally, cooperation with conservative Arab states of the region such as Saudi Arabia, Yemen, Jordan and even Egypt for resolving regional crises. On this basis, President Ahmadinejad sent his special envoys to Jordan and Yemen in the framework of enhancing Iran's active and effective role in regional issues. Another example was the replacement of Iranian Foreign Minister Manoocher Mottaki and the appointment of Ali-Akbar Salehi in his place, which was in line with the prioritizing of expanding relations with regional countries, especially with Turkey and Saudi Arabia.[84]

Third, legitimizing Iran's nuclear activities in the region. This policy was adopted by introducing a series of new principles and independent remedies aimed at gaining regional support for the legitimacy of Iranian nuclear policies: 1) Iran declared that it would support the expansion of nuclear activities in the region.[85] It even announced that it was ready to share its expertise and technical know-how with other countries in the region. 2) By offering principles of "energy security" and "nuclear justice," Iran has tried to deviate from concepts of "deterrence" and "weaponization" in its nuclear activities. Iran has

[83] "America Reacts to Ahmadinejad's Remarks about Nuclear Negotiations Nuclear" (in Farsi), *Iran Diplomatic* (Ordibehesht 20, 1390 or May 10, 2011) at http://irdiplomatic.com/view-4092-%D8%A2%D9%.html.

[84] "Interview with Foreign Minister Ali Akbar Salehi" (in Farsi), *Khabaronline* (Farvardin 2, 1390 or March 22, 2011) at http://www.khabaronline.ir/news-136326.aspx.

[85] Mehmanparast, "Iran is Ready to Cooperate."

sparked the motivation of the regional states to use peaceful nuclear energy; today, states such as Turkey, Egypt and the Persian Gulf countries have a higher interest in using nuclear energy for sustainable development. 3) Finally, Iran has revived a nuclear disarmament movement in the region. This policy has many supporters among key Middle Eastern actors including Saudi Arabia, Turkey and Egypt. By acting upon this policy, Iran intends to put pressure on the Israeli regime on the one hand, and stress its own peaceful nuclear activities on the other.

Fourth, the concept of "collective security" in the region. The prominence of Iran's role in settling regional crises, combating terrorism and drug trafficking, and defending human security in general has caused Iran to link its nuclear issue to the subject of collective security in the region in the form of a political-security package, which is to be considered from all angles simultaneously. From this perspective, threatening the use of force and adopting systematic economic sanctions on the basis of Iran's nuclear possession jeopardize the security of Iran. From Iran's perspective, the West in an unjust manner links the Iranian nuclear program with the issue of deterrence and threat to the regional and world security. Such a policy weakens Iran's stance in the nuclear talks. For example, in the last round of talks in Istanbul, the secretary of the Supreme National Security Council, Said Jalili, spoke at length about Iranian efforts to resolve regional and international problems.[86] This way, Iran establishes a link between its own nuclear issue and its regional energy production and political and human security.

Iran's nuclear program and foreign actors

Beyond the domestic and regional levels, Iran's nuclear program affects the country's relations with the world's great powers with regards to four main sets of issues. First, the issue of deterring and containing Iran is related to the preservation of the regional balance of power and to the political-security trends in the region, most notably Arab-Israeli relations. Second, the issue of supremacy and preserving the traditional authority of powers in the United Nations Security Council (UNSC) is related to maintaining great powers' monopoly over global disarmament and the adoption of UNSC sanctions. Third, the issue of preserving dominant political-legal trends in the IAEA is challenged by Iran's new model of nuclear activities in accessing an independ-

[86] "Iran, P5+1 hold talks in Istanbul," *Press TV* (January 21, 2011) at http://www.presstv.ir/detail/161206.html

ent fuel cycle and acquiring nuclear "capability" without weaponization. Fourth, the issue of international energy security and economic exchanges with Iran has also affected the policies of actors such as China and the EU.

The US: confronting actor

Since the start of the Iranian nuclear dilemma in 2002, American policies have been aimed at challenging the "independent" aspect of the Iranian nuclear program and its basic principle, which is the enrichment of uranium on Iranian soil. From the American perspective, this principle of the Iranian program could lead to the acquisition of nuclear "capability," thus paving the way toward weaponization.[87] For this reason, the US has been aggressively proactive in efforts to contain and control the Iranian nuclear program. To realize this policy, the US has adopted various strategies, including a show of force and military threat during the George W. Bush administration, a policy of "change" and direct talk with Iran during the term of Barack Obama, and now the dual policy of "sanctions for negotiations" in which the exertion of severe pressure on Iran is meant to incentivize engagement in US-led discussions and change in its nuclear policy.

The tough US stance is primarily derived on the one hand from American strategists on the right as well as the inherent US role in maintaining the Middle Eastern political-security and economic status quo, and the nation's responsibility to protect the interests of regional allies of America on the other.[88] In this respect, Iranian attempts to undertake an independent fuel cycle are coupled with the risk of weaponization. If that linkage were real, Iran's power and position in the region would increase, thus disturbing the traditional balance of power, to the detriment of the US and its allies. It is no surprise then that the US policy toward Iran's nuclear program has been tied to issues of "deterrence" and "regional security."

On the issue of "deterrence," the main objective has been to gain control over dominant political-security currents in the Middle East through containing the Iranian nuclear program. Some experts even maintain that in order to keep its global hegemony the US must dominate the Middle East, a feat that, without a check on Iran and its nuclear program, is impossible.[89]

[87] Collin Dueck and Ray Takyeh, "Iran's Nuclear Challenge," *Political Science Quarterly*, Vol. 122, No. 2 (Sunner 2007): pp. 189–205.

[88] Barzegar, "The Balance of Power in the Persian Gulf": p. 75.

[89] Chomsky, "The Iranian Threat."

The second issue of regional security, and that of Israel, has been among the top US strategic aims in the region, which would be disturbed by the independent characteristic of Iran's nuclear development.[90] With access to some 100–400 nuclear warheads[91] and the West's implicit support, Israel has been able to exploit for its own benefit the traditional deterrence doctrine in the region. The Israeli regime portrays the Iranian nuclear program as an "existential threat," connecting it with Israel's very survival as a nation.[92] Admittedly, one major stipulation of Iran's nuclear policy is the comprehensive nuclear disarmament of the Middle East, which primarily affects Israel.

Another reason for the US opposition to the Iranian nuclear program is to prevent its being tied to regional security issues.[93] For this reason, the US opposed all of Iran's proposals that placed the nuclear program in the context of regional security, as Iran elaborated during the Istanbul nuclear talks on its role in solving regional and world issues.[94]

The EU: independent and mediatory actor

For the EU, the Iranian nuclear program is important in several political-security respects. Three major actors of the EU—Germany, France, and the UK—consider economic and political-security interests in the regional and international equations. France and the UK are permanent members of the UN Security Council enjoying veto powers. Depending on its economic status and political influence, Germany also claims rights and responsibilities at the global level.

Since the onset of the Iranian nuclear issue, the role played by the EU has been under the influence of two factors: first, the US pressure on the EU, and second, the lack of a comprehensive strategy among EU members in dealing

[90] Massimo Calabresi "Iran's Nuclear Threat"; Barzegar, "The Balance of Power in the Persian Gulf": pp. 74–87.

[91] Chuck Freilich, "Decision Time in Jerusalem," *Journal of International Security Affairs*, No. 18 (Spring 2010): p. 55.

[92] Cohen, "Israel Ponders a Nuclear Iran."

[93] Farhang Jahanpour, "Iran's Nuclear Programme and Regional Security," Oxford Research Group (April 2007) at http://www.isn.ethz.ch/isn/Digital-Library/Publication/Detail/?ots591=0c54e3b3-1e9c-be1e-2c24-a6a8c7060233&lng=en&id=43705.

[94] The Istanbul nuclear talks which I speak of took place in October 2009, December 2010 and January 2011.

with Iran's nuclear issue. Affected by American pressure, which primarily emphasizes nuclear "weaponization," thereby blocking the possibility of reaching a mutual accommodation, the EU negotiations with Iran in 2004 failed. Not only did Iran lose its confidence in the EU's ability to advance nuclear talks; gradually yet inevitably, the EU also had to abandon its independent and mediatory role in the Iranian nuclear dossier, taking harder stances and eventually collaborating and coordinating with the US, thereby adopting principles of deterrence and anti-weaponization. The nuclear issue was finally pushed to the Security Council, the result of which was various sanctions resolutions including Resolution 1929. The lack of a clear and united strategy in proceeding with the nuclear talks weakened the EU's position before Iran. Since France and the UK possess nuclear weapons and played a vital role in building the structure of legal and technical rules thereof, they regard the Iranian nuclear issue through the prism of prestige, making their stance harder and closer to that of the US.[95] Germany, on the contrary, demands synergy of relations and the resumption of diplomatic negotiations, due to its higher level of interaction and economic interests with Iran.

For Iran, the importance of the EU role is in strengthening the "diplomatic approach." Contrary to the US, the EU does not demand a complete freeze of the Iranian nuclear program, and it might even be gradually convinced to recognize Iran's right to independent uranium enrichment. The EU has less political and strategic conflict with Iran and does not aspire to deny Iran its regional role. Geographically, Europe is closer to Iran and the Middle East, and its approach and policies in the region are largely predicated on resolving regional crises, preventing extremism, combating drugs, and eliminating ethnic violence.[96] The EU's failure in resolving these crises will result in the extension of tensions to Europe's borders. With an accommodating and interacting policy, the EU strives to settle these issues through cooperation with regional actors like Iran.[97] The European idea of all-out nuclear disarmament is also

[95] See Naser Sghafi Ameri, "France Policy toward Iran's Nuclear Program," strategic report, Center for Strategic Research (Tehran: August-September 2010).

[96] See Mostafa Dolatyar, "Management of Sustainable Security in the Persian Gulf, NATO's Participation: Constructive or Destructive?" (in Farsi) *Journal of Foreign Policy*, No. 1 (Spring 2010): pp. 7–22.

[97] Kristin Archick, "European Views and Policies Toward the Middle East", report for Congress by the Center for Strategic Research (Tehran: March 9, 2005)at http://www.fas.org/sgp/crs/mideast/RL31956.pdf; Kayhan Barzegar, "The European Union and Future Nuclear Talks," Belfer Center for Science and International Affairs, (Cambridge, MA: December 4, 2010).

different from the American one,[98] since geographical proximity makes any Middle Eastern nuclear arms race a serious threat to European security. Therefore, the EU (of course with some reservations over Israel's nuclear arsenals) favors comprehensive nuclear disarmament in the region.[99]

Russia and China: balancing actors

China and Russia play a "balancing" role in advancing Iran's nuclear program, because both states attribute significance to the role of the IAEA and Iran's legitimate rights in the framework of the NPT. Although Russia and China supported sanctions in Resolution 1929, Iran still demands that these two states should have a role in its nuclear program.[100] Attempting to balance regional and independent national roles as well as maintaining an international responsibility is an important point in the Chinese and Russian perspectives toward Iran's nuclear program.[101] In dealing with Iran's nuclear standoff, these two states try to act in accordance with their independent national and regional interests, as well as with their global responsibilities as permanent members of the Security Council.[102] They are driven toward Iran because they are self-interested players focused on preserving their economic and political interests at regional and international levels.[103] War with Iran would jeopardize their interests and reduce the significance of their role in the Iranian nuclear standoff.

Of course, the way Iran sees both Russia and China is different. The main issue with Russia is how it balances between "deterrence" (weaponization) and

98 Daniel Feakes, "The Emerging European Disarmament and Non-Proliferation Agenda on Chemical and Biological Weapons," *Disarmament Diplomacy*, No. 65 (July–August 2002) at http://www.acronym.org.uk/dd/dd65/65op2.htm.

99 See "Ten Myths about Iran's Nuclear Issue," paper at the British Embassy in Tehran (June 16, 2008) at http://ukiniran.fco.gov.uk/en/newsroom/?view=PressR&id=4157923.

100 Abolfazl Zohrehvand, *IRNA* (May 5, 2010).

101 See Kayhan Barzegar, "China: A Short-Term Solution for Iran," *Iran Review* (January 20, 2010) at http://www.iranreview.org/content/view/5262/36/.

102 See Mohammad Sadri, "Did Istanbul Negotiations Fail?", *Iran Review* (January 26, 2011) at http://www.iranreview.org/content/Documents/Did_Istanbul_Negotiations_Fail_.htm; Dmitri Trenin, "Russia's Nuclear Policy in the 21st Century Environment," proliferation papers, Security Studies Department, IFRI (Autumn 2005) at http://www.ifri.org/files/Securite_defense/prolif_12_Trenin.pdf.

103 See Stephen Walt, "The Sanctions Paradox," *Foreign Policy* (Washington, DC: April 21, 2010).

"peaceful" nuclear energy. Russia currently utilizes a two-pronged approach, as it is in the transition of détente with the West and is also seeking to portray itself as a powerful, independent and responsible actor at the regional and international levels. Russia has some economic, political and strategic interests in the fields of energy security, arms exchange, and nuclear activities and cooperation with Iran, the loss of which would carry a high price for Moscow.[104] At the same time, Russia wants to preserve its relations with the West in key economic and strategic areas.[105] A major concern of Russia is to keep a balance between its relations with Iran and the West. For instance, the Russians sometimes indicate a compliance with operating the nuclear power plant of Bushehr to the benefit of Iran or abstaining from joining the oil products embargo against Iran in order to demonstrate their independence and reliability with respect to economic commitments and respect for neighboring states. Simultaneously, Russia worked alongside Western actors in adopting Resolution 1929. However, Iran and Russia have common strategic needs. Being geographically close to Iran, Russia has a vested interest in cooperation on nuclear activity, energy security, natural gas and issues related to regional security in Afghanistan and the Caucasus.[106]

On the issue of deterrence, Russia undoubtedly considers weaponization of Iran's nuclear program as a threat to regional and international security and therefore cooperates with the West to prevent this scenario. Russia also considers nuclear weapons as a source of global status and prestige,[107] and is therefore not ready to accept new states into the nuclear club. Global disarmament, as Russia sees it, is a concern of the great powers and incompatible with the Iranian nuclear disarmament policy.[108] At the same time, Russia accepts the advancement of the Iranian peaceful nuclear program as long as it relies on Russian fuel and technology. The Iranian nuclear program implies major eco-

[104] Flynt Leverett and Hillary Mann, "Reading Russia on Iran," *The Race for Iran* (November 15, 2009) at http://www.raceforiran.com/reading-russia-on-iran.

[105] Vadim Kononenko, "Russia in 2008 and Beyond, What Kind of Foreign Policy?", Briefing Paper 17, The Finnish Institute of International Affairs (March 12, 2008) at http://www2.irna.com/ar/news/view/line-17/0812174879181945.htm.

[106] Flynt Leverett and Hillary Mann, "Medvedev Meets Obama: Russia-Iran Relations Should be Re-examined," *The Race for Iran* (June 2, 2010) at http://www.raceforiran.com.

[107] Trenin, "Russia's Nuclear Policy in the 21st Century Environment."

[108] Kayhan Barzegar, "The Win-Win Strategy in Iran's Nuclear Program," *Iranian Diplomacy* (January 1, 2011).

nomic and strategic interests for Russia, the most important of which are: 1) return of the Bushehr nuclear reactor's spent fuel; 2) annual charges of the Bushehr nuclear power plant and monitoring safety regulations; 3) inspection and control of the plant for which the Russian company AtomStoryExport is responsible; 4) training Iranian engineers and technicians; 5) building new plants for Iran; and finally, 6) acquiring regional markets for nuclear cooperation, which in turn stabilizes the Russian stance in the future of the regional and international nuclear industry.[109] Turkey and Egypt are two potential clients, as Turkey already signed a cooperation agreement with Russia in 2004[110] and Egypt has also made cooperation deals with Russia in the past.[111]

However, Iran's stance on China differs from the stance on Russia, since Iran expects China to play a balancing role in advancing the Iranian nuclear program. China has the mildest and the most positive view of the Iranian nuclear program, trying to reconcile its short-term and long-term interests. China's short-term interests with Iran are based on the security of energy flow, expansion of economic ties, a peaceful settlement of the Iranian nuclear standoff, and warding off the risk posed by war in the region.[112] In the long term, China seeks good relations with the West, especially the US, and with the international institutions that preserve traditional arrangements of authority of the Security Council. In this respect, Chinese and Iranian interests are at least aligned on one common point—favoring diplomacy and opposing military action or unilateral and tough sanctions against Iran, which would, from China's perspective, challenge the Security Council's authority.[113] China

[109] Kayhan Barzegar, "Russia and the Future of Nuclear Talks," *Iran Review* (August 25, 2010) at http://www.iranreview.org/content/Documents/Russia_and_the_Future_of_Nuclear_Talks.htm; Rajab Safarof, "The Importance of the Bushehr Plant for Russia," *Tabnak* (Shahrivar 2, 1389 or August 24, 2010) at http://www.tabnak.ir/fa/news/116177.

[110] Saban Kardas, "Russia will construct Turkey's first nuclear power plant," *Eurasia Daily Monitor*, Vol. 7, No. 98 (May 20, 2010) at http;//www.jamestown.org; "Medvedev Signs Russia-Turkey Nuclear Papers," *Ria Novosti* (May 7, 2011) at http://en.rian.ru/russia/20110507/163909280.html.

[111] "Russia-Egypt nuclear deal signed," *BBC News* (March 25, 2008) at http://news.bbc.co.uk/2/hi/europe/7313037.stm.

[112] Stephen Walt, "I'll Believe It When I See It...," *Foreign Policy* (Washington, DC: April 13, 2010) at http://walt.foreignpolicy.com/posts/2010/04/13/ill_believe it when i see it.

[113] Paul Richter, "U.S. and EU fail to isolate Iran," *Los Angeles Times* (August 8, 2010) at http://www.articles.latimes.com/2010/.../la-fg-iran-sanctions-20100809.

recently extended the annual contract for the importation of crude oil from Iran.[114] With regard to deterrence and the possibility of the weaponization of the Iranian nuclear program, China does not consider the Iranian nuclear program a threat to its national interests and security,[115] and holds that Iran has the full right to enrich uranium and have access to peaceful nuclear energy.[116] And also contrary to Russia, the US, and some European countries like France, China favors a global zero policy and thus does not view the Iranian nuclear program through the exclusive prism of deterrence and proliferation.[117] Instead, China's nuclear stance is very close to Iran's, demanding a comprehensive nuclear disarmament in the world and in the Middle East.[118] China's appreciation of the International Conference on Disarmament held in Tehran is proof of this.[119]

Turkey and Brazil: new actors

For Iran, the emergence of new actors who understand Iran's motives, and are closer strategically to its nuclear policies, creates a new confidence-building environment and strengthens Iran's bargaining power in nuclear discussions. From Iran's perspective, Turkey and Brazil have defined their role as neutral intermediaries not harboring hegemonic ambitions. On the other hand, these rising powers demand a strengthening of their regional and international roles depending on their own independent, domestic strategies.[120] Their views are

114 "China Extended the Contract of Crude Oil Imports From Iran in 2011" (in Farsi), *Aftab News* (Bahman 1, 1389 or January 21, 2011) at http://www.aftabnews.ir/vdchvwnzz23nz6d.tft2.html.

115 See Matthew Kroening, "Bombs Away," *The New Republic* (Washington, DC: February 9, 2010).

116 Naser Saghafi Ameri, "China and Iran's Nuclear Program," strategic reports, Center for Strategic Research(Tehran: August-September 2006); George Perkovich, "Five Scenarios for the Iranian Crisis," proliferation papers, Security Studies Department, IFRI (Winter 2006) at http://www.carnegieendowment.org/files/Prolif_Paper_Perkovich_Iran_Scenarios.pdf.

117 Ming Zhang, *China's Changing Nuclear Posture* (Washington, DC: Carnegie Endowment for International Peace, 1999).

118 "White Paper: China's National Defense," paper at the Information Office of the State Council of the People's Republic of China (2008).

119 "Iran is Pacifism, China Has Opposed Sanctions International Against Iran," *Iranian Students' News Agency (ISNA)*, (Farvardin 29, 1389 or April 19, 2010) at http://isna.ir/isna/NewsView.aspx?ID=News-1521155.

120 See Graham E. Fuller, "Former CIA officer on Iran: Brazil and Turkey are vital checks and balances," *Christian Science Monitor* (Boston, MA: May 24, 2010).

supported by many worldwide, especially the Non-Aligned Movement (NAM) members, developing states and Islamic countries.[121] They have already benefited from the signing of the "Tehran Declaration" on May 17, 2010, in which the transfer of nuclear fuel to Turkey was confirmed. Turkey especially took the utmost advantage of the vacuum created by the EU's policy realignment, thereby becoming a trustworthy replacement for the role of intermediary. Iran has called for Turkish and Brazilian involvement in its nuclear talks, which, of course, was opposed by the P5+1 group (the five permanent members of the UNSC plus Germany, joined in their diplomatic efforts with Iran in regards to its nuclear program).[122] However, Turkey's importance is signified by Istanbul's hosting of the last round of talks in January 2011.

Iran and Turkey have common positions on three issues, which lead them to complement one another's roles in the Iranian nuclear standoff. First, Turkey has never supported the West's sanctions policy and the pursuit of military means against Iran,[123] as sanctions would inflict heavy economic losses upon the Turkish economy as well.[124] The volume of exchanges is supposed to increase from the existing level of US$10 billion per annum to US$30 billion.[125] Turkey has announced its opposition to unilateral sanctions and an oil

[121] "Iran is Committed to NPT" (in Farsi), *Mehr News* (Bahman 3, 1389 or January 23, 2011) at http://www.mehrnews.com/fa/newsdetail.aspx?NewsID=1238251; Ali Asghar Soltanieh, "NAM supports Iran's Nuclear Activities," *English IRIB* (March 5, 2011) at http://english.irib.ir/news/political/item/71850-nam-supports-irans-nuclear-activities; "Malaysian FM says Malaysia acknowledges Iran's nuclear rights," *IRNA* (May 26, 2011) at www.irna.ir/ENNewsShow.aspx?NID=30404957.

[122] Robert Dreyfus, "United States slams Turkey, Brazil over Iran," *The Nation* (New York: May 28, 2010).

[123] "Erdogan: Diplomacy should settle Iran's nuclear question," *Hurriyet* (April 6, 2010); "The Chancellor of Germany, in Ankara, said: 'Our view is that sanctions are not a good way ... we think the best way is diplomacy'" (in Farsi), *Farda News* (Farvardin 10, 1389 or March 30, 2010) at http://www.fardanews. com/fa/news/106807/.

[124] See İsmail Duman, "Turkey-Iran relations: friends or foes?", *World Bulletin* (November 14, 2010) at http://turkey-news.net/turkey-iran-relations-friends-or-foes/.

[125] Saban Kardas, "Turkish-Iranian economic ties flourish," *Eurasia Daily Monitor*, Vol. 8, No. 35 (February 18, 2011) at http://www.jamestown.org; Mahmoud Heydari, "Iran-Turkey Level of Economic Exchangers Increase to $ 30 Billion" (in Farsi), *Iran Economist* (Khordad 2, 1389 or May 23, 2010) at http://ww w.iraneconomist.com/economic/ecnomy-bazzar/26261—30—.html.S.

products embargo. War also would disturb other political-security interests of Turkey, as the 2003 invasion of Iraq has already proved. Therefore, the outbreak of another war in the region would be disastrous for the Turkish economy, which has recently enjoyed a steady growth rate.[126] Second, Turkey, like Iran has called for the implementation of all-out nuclear disarmament in the Middle East.[127] Both are opposed to the West's double standards in dealing with the Israeli nuclear arsenal and Israel's refusal to sign the NPT. Both states consider nuclear weapons a threat in their national security strategy.[128] Third, there is the issue of future nuclear cooperation between Turkey and Iran. Turkey, like Iran, aims to use nuclear energy for sustainable development for which, as mentioned earlier, Turkey signed a nuclear agreement with Russia in 2004 in order to start the construction of nuclear reactors. One of the main reasons why Turkey supports the Iranian nuclear program is its own effort and willingness to make use of nuclear energy.[129]

Conclusion

This chapter has examined how a long precedent of nuclear activities—simultaneously attaining different economic, technological, development, identity and political goals—turned the Iranian nuclear program into a national and strategic issue for the country. It has also discussed how there is correlation and interdependence between different aspects of domestic, regional and international politics dimensions of Iran's nuclear program. Domestically, the most important principle in the Iranian nuclear program, "uranium enrichment" on Iranian soil, has created a near political consensus among the political-security elites toward direct and comprehensive talks with the US in the regional and international spheres, and toward prioritizing the strategic and national security issues in the country's foreign policy.

At the regional level also, the Iranian nuclear program is interdependent on regional politics. This is due to the nature of the Iranian nuclear program,

126 Ben Katcher, "Turkey Foreign Minister Ahmet Davutoglu on Iran's Nuclear Program," *The Race for Iran* (April 15, 2010) at http://www.theraceforiran.com.

127 Scott D. Sagan, "Shared Responsibilities for Nuclear Disarmament," *Daedalus*, Vol. 138, No. 4 (Fall 2009): pp. 157–68.

128 Prime Minister of the Republic of Turkey, Rajab Tayeb Erdogan, "Turkey Criticized the Dual Approach of the West about the Nuclear Issue" (in Farsi), *BBC Persian* (Khordad 9, 1389 or May 30, 2010) at http://www.bbc.co.uk/persian/iran/2010/05/100529_u02-erdogan-iran-us-nuclear.shtml?print=1?print.

129 "Iran's Nuclear Program: The Way Out," *Time* (New York: May 9, 2006).

which is connected to the economic, security and political issues in the region and, as a consequence, to relations with the regional and trans-regional players. From this perspective, the Iranian nuclear program has increased the desire of regional states to make use of peaceful nuclear energy for the purposes of sustainable development and economic growth. Meanwhile, the Iranian nuclear program stands in direct relation to deceleration or acceleration of the arms race and to the preservation of the traditional regional balance imposed by foreign powers. The second subject at the regional level is the regionalization of Iran's nuclear program. This Iranian policy is aimed at tying its security to that of the Middle East, and the world at large, through the nuclear issue. It attempts this by creating "strategic political parity" in the nuclear talks with the West, legitimizing peaceful nuclear activities at the regional level by supporting the expansion of the nuclear activities of other regional states, and proposing collective and comprehensive security measures in regional issues.

At the international level, the interdependence between the policies of trans-regional actors in world issues and the NPT, and their approaches in dealing with the Iranian nuclear program, were here also examined. The US was introduced as a state that views the Iranian nuclear program as one based on deterrence and as a threat to its national security. The US has also become the main proponent of bringing Iran's nuclear program to a halt. The EU (Germany, France and the UK), which mainly views the Iranian nuclear program through the prism of non-proliferation, is considered by Iran as an intermediary actor with a view to strengthening the diplomatic approach. Russia and China, which rely on strengthening the legal status of the IAEA on the Iranian nuclear issue, are considered balancing actors in promoting the Iranian nuclear program. Finally, Turkey and Brazil are taken by Iran to be new actors in the field of diplomacy, who stress the legitimacy of peaceful Iranian nuclear activities.

The main challenge facing Iran and the West in the nuclear standoff is how to strike the appropriate balance between Iran's right to the peaceful use of nuclear energy guaranteed under the NPT and the possibility of nuclear weaponization. This standoff will only be settled by working toward a win-win outcome: recognition of the right to enrichment on Iranian soil by the West (a win for Iran) and provision of the necessary assurances on the side of Iran for non-diversion (a win for the West). However, the existing mistrust between the two sides, which itself is bound up in political and ideological differences, is deep-rooted and continues to be a serious impediment to reaching a sustainable solution. Despite the existing challenge, the Iranian nuclear program has created the first serious opportunity to engage in direct and strategic negotiations with the West, and especially the US. This indicates that Iran's nuclear program has led the two sides to approximate their mutual appeals and to prevent crises, which are capable of endangering regional and international security. Entering the fourth round of direct Geneva talks is a good example of this.

Another issue relating to the Iranian nuclear program is that of problems resulting from a possible agreement between Iran and the West. Recognizing Iran as a nuclear state means accepting the Iranian model of nuclearization within the framework of the NPT. This development would itself have a real impact on issues related to global and regional politics. By recognizing a nuclear Iran, new initiatives on the part of the West would mainly consider how to control and manage the Iranian nuclear program, resorting to a modified Additional Protocol, which increases the probability of controls and inspections being mitigated, and overall it would be confined to preventing the shift from nuclear "capability" to "weaponization." From the other side, issues related to the comprehensive nuclear disarmament in the region and pressuring the Israeli regime to accept the NPT would enter a new stage in regional relations. Finally, the Iranian nuclear program has brought the demand for nuclear energy in the Middle East into a new phase. After Russia tried to keep its monopoly over the Iranian nuclear market, Western states like the US, France, Canada and Germany, and their Asian allies like Japan and South Korea, entered the stage as suppliers of nuclear energy to the Middle East. In the future, countries considering access to nuclear "capability" and an "independent nuclear fuel cycle" might instead follow Iran's example. Here, the main challenge for the West is to choose between accepting the movement of states toward acquiring such capability and their complete dependence on Western technology.

10

A Nuclear Weapons Free Zone in the Middle East

Gawdat Bahgat

The founding of Israel in 1948 drastically altered the Middle Eastern landscape, particularly the security dynamics. Initially, Arab countries rejected the existence of a Jewish state at the heart of the Middle East. The 1948, 1956, 1967, and 1973 wars, as well as several other conflicts, can be seen as clear signs and demonstrations of this deep-rooted mutual hostility between the two sides. The Arab-Israeli conflict took a turning point in the late 1970s with the signing of the Egyptian-Israeli Peace Treaty in 1979. More than ten years later, Jordan signed a similar treaty and the Palestinian Liberation Organization (PLO) signed the Oslo Agreement. Meanwhile, other Arab countries, particularly in the Gulf, engaged in commercial relations with Israel without explicitly awarding diplomatic recognition. These developments have reflected a growing realization by both Arab governments and the "Arab Street" that Israel is there to stay and has become an undisputed part of the region's landscape. In short, it can be argued, in recent decades the core of the Arab-Israeli conflict is less about the mere existence and legitimacy of Israel and more about liberating territories occupied in the 1967 war, the return of Palestinian refugees, and the establishment of a Palestinian state with east Jerusalem as its capital.

This mutual acceptance, however, has been challenged by the lack of consensus on how to address the nuclear weapons issue. Since the late 1960s to early 1970s Israel has been widely believed to possess nuclear weapons. As the only nuclear power in the Middle East, Israeli governments have strictly adhered to a policy known as the Begin Doctrine under which Jerusalem has vowed not to allow its neighbors to develop nuclear weapons. Guided by this policy, the Israeli Air Force carried out two successful strikes on the Iraqi nuclear reactors (1981) and the Syrian nuclear plant (2007). In addition, Israel is closely watching the development of Iran's nuclear program and has repeatedly threatened to take military action to destroy Tehran's nuclear capabilities.

On the other side, Israel's nuclear monopoly has left the Arabs and Iranians with a sense of vulnerability and inferiority. Little wonder that some of Israel's neighbors have sought to acquire nuclear, biological and chemical weapons. The goal is to reach a state of balance of power between the two sides. Arab and Iranian efforts to acquire nuclear capabilities have not succeeded and Israel remains the only nuclear power in the Middle East. Another approach to counter Israel's nuclear arsenal is to pressure Jerusalem to give up its nuclear weapons and establish a nuclear weapons free zone (NWFZ) in the Middle East. Under such a scheme all regional powers will not have access to nuclear weapons and the nuclear military balance between Israel and its neighbors will be restored.

This chapter seeks to examine the prospects of establishing a NWFZ in the Middle East. In the next section I provide a definition of the concept of a nuclear weapons free zone, the roots of this concept, and how it is related to the Non-proliferation Treaty (NPT). This will be followed by a close examination of other regions' experiments with a NWFZ. Specifically, I discuss the five NWFZs already established. Then the analysis focuses on the efforts to reach a consensus on creating a similar zone in the Middle East and the opposing Arab-Iranian and Israeli approaches. I also discuss the potential impact of the growing interest in peaceful nuclear power in Iran and several Arab countries on the proliferation of nuclear weapons. In the concluding section I summarize the main findings of the study and explore the potential for ridding the Middle East of nuclear weapons. I argue that the efforts to create a NWFZ in the Middle East should be part of a broader strategy to slow the conventional arms race and reach a genuine and comprehensive peace agreement. The intense unrest in several Arab countries since early 2011 adds more urgency to reaching a consensus on a regional security system.

Nuclear weapons free zone: definition

A NWFZ is a regional arrangement that prohibits the "development, manufacture, stockpiling, acquisition, possession, control, along with assistance in research on the development, manufacture, stockpiling or acquisition, or possession of any nuclear explosive device within the zone of application by any contracting party."[1] By creating geographic areas that are completely free of nuclear weapons, these zones contribute to world peace and to making the entire world free of these weapons systems. Indeed, NWFZs and the NPT pursue similar goals. While the NPT takes a global approach to nuclear non-proliferation and disarmament, NWFZs complement this approach on a regional basis. They further strengthen the non-proliferation and disarmament norm, and reduce the geographic area in which nuclear weapons can be deployed or used. The NPT recognizes this role of NWFZs. Article VII of the NPT states that "nothing in this Treaty affects the right of any group of States to conclude regional treaties in order to assure the total absence of nuclear weapons in their respective territories."[2]

United Nations General Assembly (UNGA) Resolution 3472 (1975) states that as a general rule, a NWFZ should "be deemed to be any zone recognized as such by the United Nations General Assembly, which any groups of states, in the free exercise of their sovereignty, have established by virtue of a treaty or convention."[3] According to the Resolution the treaties that establish a NWFZ have to meet two conditions: 1) a total absence of nuclear weapons, including the procedure for the delimitation of the zone; 2) an international system of verification and control to guarantee compliance with the stated obligations.

The states party to the NWFZ treaties usually receive legally binding security assurances from the five NPT nuclear weapons states (China, France, Russia, the United Kingdom, and the United States), who, by signing and ratifying the relevant protocols to the treaties, pledge to respect the status of NWFZs and not to use or threaten to use nuclear weapons against states party to these treaties. Such declarations of non-use of nuclear weapons are referred

[1] NWFZ Clearinghouse, Center for Nonproliferation Studies, at http://cns.miis.edu/nwfz_clearinghouse/index.htm (last accessed May 14, 2008).

[2] United Nations, Non-Proliferation Treaty, at http://www.un.org/Depts/dda/WMD/Treaty (last accessed April 5, 2011).

[3] United Nations General Assembly, Resolution 3472 (1975) at http://www.opanal.org/docs/un/unag30res3472i.pdf (last accessed April 5, 2011).

to as negative security assurances. However, the five nuclear-armed countries have at times signed and ratified a NWFZ protocol and declared conditions reserving the right to use nuclear weapons in certain scenarios against parties to a NWFZ. For instance, the United States signed the protocol for the African NWFZ in 1996 with a declaration that it would reserve the right to respond with all options, implying possible use of nuclear weapons, to a chemical or biological weapons attack by a member of the zone. Similarly, none of the nuclear weapons states have signed the relevant protocol for the treaty creating a zone in Southeast Asia because of concerns that it conflicts with the right of their ships and aircraft to have freedom of movement in international waters and airspace.

In addition to contributing to regional and global disarmament, NWFZs facilitate confidence building among regional powers and improve the chances of cooperation in the peaceful use of nuclear technology. Other potential areas of cooperation include the prohibition of dumping of radioactive waste in the oceans and the establishment of safety standards against the theft of nuclear materials.

The concept of a NWFZ has its roots in the Cold War and the fear of a nuclear confrontation between the superpowers of the US and the Soviet Union. In 1956 the Soviet Union introduced the idea of a NWFZ in Central Europe at the UNGA. Two years later, in 1958, the Polish government proposed a NWFZ covering Poland, Czechoslovakia, both German states, and other European countries. The Polish government was mainly concerned about Americans deploying nuclear weapons in West Germany and the Soviets deploying nuclear weapons in Poland. In addition to the banning of the manufacture, possession, stationing, and stockpiling of nuclear weapons and equipment, the proposal called for the prohibition of nuclear attacks against state members in the zone. Finally, the plan included the establishment of an international verification mechanism.[4] Neither the Soviet suggestion nor the Polish proposal (named the Rapacki Plan after the Polish foreign minister, Adam Rapacki) made any progress due to the mutual suspicion and hostility that characterized the Cold War era. In the ensuing years other countries including Sweden, Finland, and Romania made similar proposals. These initiatives often stemmed from the desire to ensure peace in Europe and prevent a nuclear holocaust. No wonder, the Cuban Missile Crisis in 1962 (the closest

[4] Nuclear Threat Initiative, *Nuclear Weapons Free Zones*, at http://www.nti.org/h_learnmore/nwfztutorial/chapter02_03.html (last accessed April 5, 2011).

the world had ever come to a nuclear confrontation) served as a catalyst for the creation of NWFZs in Latin America and elsewhere.

The global setting

There are five NWFZs in the world, created by five regional agreements: the Latin and Caribbean NWFZ (the 1967 Treaty of Tlatelolco), the South Pacific NWFZ (the 1985 Treaty of Rarotonga), the Southeast Asian NWFZ (the 1995 Treaty of Bangkok), the African NWFZ (the 1995 Treaty of Pelindaba), and the Central Asian NWFZ (the 1997 Treaty of Semipalatinsk). One hundred and eleven countries with approximately two billion people are states party to these NWFZs.[5] In addition, there are treaties banning the deployment of nuclear weapons in Antarctica, Mongolia, on the seabed, and in outer space. Thus, NWFZs are "part and parcel of the global nuclear non-proliferation regime."[6]

Different dynamics created the appropriate conditions for establishing each NWFZ. As Muhammad ElBaradei, former Director General of the International Atomic Energy Agency (IAEA), explains, "Because the causes of insecurity vary from region to region, security solutions do not come in a 'one-size-fits-all' package. It is for this reason that regional dialogues and NWFZs are so beneficial."[7]

The African Nuclear Weapon Free Zone Treaty (Treaty of Pelindaba)

In 1961 the UNGA adopted a resolution calling on Member States to consider and respect the continent of Africa as a de-nuclearized zone. In 1964 the Organization of African Unity issued the Declaration on the de-nuclearization of Africa which was subsequently endorsed by the UNGA.[8] The Treaty

[5] Wakana Mukai, "The Importance of Nuclear Weapons Free Zones," *Journal of Science and World Affairs*, Vol. 1, No. 2 (June 2005): pp. 79–86, p. 81.

[6] Jozef Goldblat, "Nuclear-Weapon-Free Zones: A History and Assessment," *Nonproliferation Review*, Vol. 4, No. 3 (Spring-Summer 1997): pp. 18–32, p. 19.

[7] Muhammad ElBaradei, "Nuclear Weapons Free Zone: Pursuing Security, Region by Region," Statements of the Director General, Conference of States Parties and Signatories of Treaties that Establish Nuclear Weapons Free Zones, Tlatelolco, Mexico (April 26, 2005) at http://www.iaea.org/NewsCenter/Statements/2005/ebsp2005n005.html (last accessed January 24, 2006).

[8] Nuclear Threat Initiative, African Nuclear Weapon Free Zone Treaty, at http://www.nti.org/e_research/official_docs/inventory/pdfs/anwfz.pdf (last accessed April 16, 2011).

prohibits the research, development, testing, possession, control or stationing of nuclear explosive devices in the territory of parties to the Treaty and the dumping of radioactive wastes. The Treaty also prohibits any attack against nuclear installations in the zone by Treaty parties and requires them to maintain the highest standards of peaceful purposes. The Treaty requires all parties to apply full-scope IAEA safeguards to all their peaceful nuclear activities. The Treaty established the African Commission on Nuclear Energy to serve as the mechanism to verify compliance.[9] The treaty was signed in 1996 and came into effect with the twenty-eighth ratification on July 15, 2009.

The South Pacific Nuclear Free Zone (SPNFZ) Treaty (Treaty of Rarotonga)

The Treaty was signed by the South Pacific nations of Australia, the Cook Islands, Fiji, Kiribati, Nauru, New Zealand, Niue, Papua New Guinea, the Solomon Islands, Tonga, Tuvalu, Vanuatu, and Western Samoa on the island of Rarotonga on August 6, 1985, and has since been ratified by all of those states. Several islands in the Indian Ocean also belong to Australia and are therefore part of the zone.[10] The Treaty bans the manufacture, possession, stationing and testing of any nuclear explosive device. It also bans the dumping of radioactive wastes at sea.[11]

The Southeast Asian Nuclear Weapon Free Zone (SEANWFZ) Treaty (Bangkok Treaty)

This is based on a notion dating back to November 1971, when the original five members of the Association of Southeast Asian Nations (ASEAN) signed in Kuala Lumpur a Declaration on a Zone of Peace, Freedom and Neutrality. However, due to the unfavorable political environment in the region, the formal proposal for the establishment of such a zone was tabled in the mid-

[9] Federation of American Scientists, African Nuclear Weapon Free Zone Treaty, at http://www.fas.org/nuke/control/anwfz/index.html (last accessed April 16, 2011).

[10] Nuclear Threat Initiative, South Pacific Nuclear Free Zone, Treaty of Rarotonga, at http://www.nti.org/e_research/official_docs/inventory/pdfs/spnfz.pdf (last accessed April 16, 2011).

[11] Federation of American Scientists, South Pacific Nuclear Free Zone, Treaty of Rarotonga, at http://www.fas.org/nuke/control/spnfz/index.html (last accessed April 16, 2011).

1980s. After a decade of negotiations the Treaty was signed by ten Asian countries under the auspices of ASEAN: Brunei Darussalam, Cambodia, Indonesia, Laos, Malaysia, Myanmar, Philippines, Singapore, Thailand and Vietnam. It entered into force on March 28, 1997.[12] It obliges its members not to develop, manufacture, acquire, possess or have control over nuclear weapons. It also obliges them not to provide the source of special fissionable materials or equipment to any non-nuclear weapons state unless subject to safeguard agreements with the IAEA.[13]

The Central Asian Nuclear Weapon Free Zone (CANWFZ)

This originates from the 1992 initiative by Mongolia declaring itself a NWFZ and calling for a regional NWFZ. The first formal CANWFZ proposal was made by Uzbek President Islam Karimov at the forty-eighth session of the UNGA in 1993. Additional proposals followed in 1994–6 but none made any headway due to a lack of regional consensus. The crucial step was taken in February 1997 when the five presidents of the Central Asian States (Kazakhstan, Kyrgyzstan, Tajikistan, Turkmenistan and Uzbekistan) issued the Almaty Declaration endorsing the creation of a CANWFZ. Experts from the five Central Asian States agreed on the text of a treaty at a meeting held in Samarkand in Uzbekistan in September 2002. In February 2005 they adopted a final draft of the treaty in Tashkent, Uzbekistan. The Treaty opened for signature in September 2006 and entered into force in March 2009.[14]

The five signatories agreed not to research, develop, manufacture, stockpile, acquire, or possess nuclear weapons or other nuclear explosive devices. The Treaty bans the receipt, storage, installation or other form of possession of any nuclear weapon on the territory of the member states. Each party pledges not to carry out nuclear weapon tests and pledges to prevent nuclear explosions at any place under its control. Finally the Treaty does not affect the rights and

[12] Nuclear Threat Initiative, Southeast Asian Nuclear Weapon Free Zone, Bangkok Treaty, at http://www.nti.org/e_research/official_docs/inventory/pdfs/seanwfz.pdf (last accessed April 17, 2011).

[13] Federation of American Scientists, Southeast Asian Nuclear Weapon Free Zone, Bangkok Treaty, at http://www.fas.org/nuke/control/seanwfz/index.html (last accessed April 17, 2011).

[14] Nuclear Threat Initiative, Central Asian Nuclear Weapon Free Zone, at http://www.nti.org/e_research/official_docs/inventory/pdfs/canwfz.pdf (last accessed April 17, 2011).

obligations of the parties under other international treaties concluded prior to the entry into force of the CANWFZ.[15]

The Latin and Caribbean Nuclear Weapon Free Zone (Treaty of Tlatelolco)

Costa Rica was the first regional state to propose a Latin American nuclear arms control arrangement at an Organization of American States (OAS) council meeting in 1958. Other proposals were unsuccessfully floated within the OAS context during the years 1958–60. But the regional consensus was building, and by 1963 Bolivia, Brazil, Chile, Ecuador and Mexico called for the establishment of a NWFZ in Latin America and the Caribbean. The Treaty was drafted five years after the Cuban Missile Crisis and opened for signature in February 1967 and entered into force in April 1969. Cuba was the last country to sign (1995) and ratify (2002) the Treaty, completing the signature and ratification by all 33 nations of Latin American and the Caribbean.[16]

The Treaty aims to prohibit the testing, use, manufacture, production, or acquisition of nuclear weapons directly or indirectly. Signatory states pledged to refrain from engaging, authorizing, or participating in the testing, use, production, or possession of nuclear weapons. This was the first time such a ban on nuclear weapons was imposed in a densely populated area. Praising the Treaty, the IAEA former Director General Mohamed ElBaradei stated, "the Treaty set an important precedent in devaluing the role of nuclear weapons in its zone of application, thereby contributing to regional peace and security by ensuring that Latin America and the Caribbean remained free from nuclear weapons."[17]

These five NWFZs suggest that two crucial criteria are necessary for successfully establishing a NWFZ: a common historical understanding among regional states and a manageable relationship with the five recognized nuclear weapons states. Stated differently, deep-rooted hostilities among regional

[15] Federation of American Scientists, Central Asian Nuclear Weapon Free Zone, at http://www.fas.org/nuke/control/canwfz/index.html (last accessed April 17, 2011).

[16] Nuclear Threat Initiative, Treaty of Tlatelolco, at http://www.nti.org/e-research/official_docs/inventory/pdfs/Tlatelolco.pdf (last accessed April 18, 2011).

[17] International Atomic Energy Agency, "Tlatelolco Treaty A Trailblazer for Non-Proliferation," at http://www.iaea.org/newscenter/news/2007/Tlatelolco.htm (last accessed April 18, 2011).

states and intense tensions with one or more of the nuclear weapons states are factors that are likely to complicate the creation of a NWFZ. This largely explains the failure to establish a NWFZ in the Middle East. The decades-long animosity between Israel on one side and Iran and some of the Arab states and non-state actors on the other needs to be addressed. At the same time the deep-rooted hostility between Iran and the United States needs to be resolved. The challenge, as the analysis below suggests, is which comes first, taking steps to build confidence between regional and international rivals that might lead to the establishment of a NWFZ in the Middle East or starting by establishing such a zone as a major step to pursue regional peace. For about four decades regional powers have failed to reach an agreement on how to address this challenge.

A Nuclear weapons free zone in the Middle East

The efforts to establish a NWFZ in the Middle East have been pursued at national, regional and international levels for almost four decades. Since Israel was thought to have acquired nuclear weapons capability in the late 1960s and early 1970s, its Arab and Iranian neighbors have sought to restore the balance of power by exerting diplomatic pressure on the Jewish state to give up these presumed weapons. As early as 1974 Iran under the Shah and Egypt under Sadat proposed a NWFZ in the Middle East to the United Nations General Assembly. Those two countries did not and still do not possess nuclear weapons. Their common aim was to compel Israel to dismantle its nuclear weapons. Six years later (in 1980), Jerusalem produced its own proposal and called for direct negotiations between regional powers instead of installing a zone by the international community. Arab, Iranian and Israeli diplomatic maneuvers failed to achieve any success. Meanwhile, the Iran-Iraq War (1980–8) underscored the security challenges the entire Middle East encountered.

This eight-year war between the two Gulf powers gave a momentum to the proliferation of chemical and biological weapons as well as short and medium -range missiles. This new security environment represented a new threat not only to the Persian Gulf nations, but also to Israel. In order to address this threat Egypt, the United States and the United Nations pursued different strategies.

In April 1990 the Egyptian President Hosni Mubarak proposed the establishment of a weapons of mass destruction free zone (WMDFZ) in the Middle East. His proposal comprised three components: first, all WMD—

nuclear, chemical, biological, or otherwise—should be prohibited. Second, all regional powers should make equal and reciprocal commitments. Third, verification measures and modalities should be established to ascertain full compliance.[18]

Following the liberation of Kuwait from the Iraqi occupation (1990–1) the United States made good on its promise to its Arab partners to reactivate its efforts toward a peaceful solution to the Arab-Israeli conflict. A peace conference was held in Madrid, in which Israel and most of its Arab neighbors participated. The United States, the European Union, and the Soviet Union also were represented. The participants created five multilateral working groups. Thirteen Arab states, Israel, a Palestinian delegation, and a number of extra-regional entities participated in these working groups. The focus was on confidence building. The Arms Control and Regional Security (ACRS) negotiated issues related to disarmament. The participants sought to follow the East-West experience.[19] These unprecedented talks, however, failed due to the large and deep differences between the parties, notably Egypt and Israel. The two sides could not agree on the connection between nuclear disarmament, conventional weapons, and peace.[20]

Similar efforts were made at the United Nations to promote the idea of a NWFZ in the Middle East. Article number 14 of the United Nations Security Council Resolution 687, which terminated the Persian Gulf War in 1991, states "Actions to be taken by Iraq represent steps toward the goal of establishing in the Middle East a zone free from weapons of mass destruction and all missiles for their delivery and the objective of a global ban on chemical weapons."[21] In 1998 the UN Secretary General commissioned a study on the effective and verifiable measures that would facilitate the establishment of a NWFZ in the Middle East. The report did not propose explicit language for a treaty, but did suggest a catalogue of measures to serve as confidence-build-

[18] Mohamed I. Shaker, "The Middle East Issue: Possibilities of a Nuclear Weapons Free Zone," at http://www.opanal.org/articles/aniv-30/shaker.htm (last accessed April 6, 2006).

[19] Claudia Baumgart and Harald Muller, "A Nuclear Weapons Free Zone in the Middle East: A Pie in the Sky?", *Washington Quarterly*, Vol. 28, No. 1 (Winter 2004–5): p. 47.

[20] "Arms Control and Regional Security in the Middle East", Center for Nonproliferation Studies, at http://cns.miis.edu (last accessed April 18, 2011).

[21] United Nations Security Council, Resolution 687 (1991), at http://www.un.org/docs/scres/1991/scres91.htm (last accessed April 18, 2011).

ing measures and as steps to prepare for a regime that would finally become a NWFZ.[22]

The NPT review conferences echoed similar sentiments. In 1995 the participants issued a resolution on the Middle East calling for the establishment of a NWFZ and recommending that states not party to the NPT accede to it and thereby accept an international legally binding commitment not to acquire nuclear weapons.[23] The 2010 NPT Review Conference went one step further by calling for a conference on the creation of a WMDFZ in the Middle East by 2012. Finally, the leaders of 43 countries in Europe, the Middle East and northern Africa participating in the first Mediterranean summit in July 2008 adopted a declaration calling for the creation of a verifiable WMDFZ in the Middle East.

These calls by several international forums have not succeeded in bridging the huge gap between regional powers. The lack of any progress in the Arab-Israeli peace process and the Iranian leaders' rhetoric regarding Israel have made reaching a common stand on a NWFZ in the Middle East much harder.

The Israeli approach

Israel has always adopted a skeptical view of global arms control and disarmament treaties,[24] and Israeli leaders have always stressed that the proliferation of WMD in the Middle East will have to be handled within a regional framework.[25] Several characters can be identified in the Israeli stand on a NWFZ. First, the creation of the State of Israel followed the extraordinary experience of the Holocaust in which the Nazis killed millions of Jews. This dramatic experience shaped the Israeli collective psyche, particularly in the first decades after the formation of Israel. Israeli leaders have always believed that nuclear weapons will shield them from a future Holocaust and see nuclear weapons as a last line of defense—an "insurance policy" to guarantee their survival.

[22] "Arms Control and Regional Security in the Middle East", Center for Nonproliferation Studies.

[23] Federation of American Scientists, Resolution on the Middle East, at http://www.fas.org/nuke/control/npt/text/resoluti.htm (last accessed April 7, 2006).

[24] Gerald M. Steinberg, "Middle East Arms Control and Regional Security," *Survival*, No. 36, No. 1 (January 1994): pp. 126–41.

[25] Emily Landau and Tamar Malz, "Israel's Arms Control Agenda," *Strategic Assessment*, Vol. 2, No. 4 (March 2000) at http://www.tau.ac.il/jcss/sa/index.html (last accessed March 15, 2000).

Refusal to recognize Israel and rhetoric calling for its destruction only feed this belief in the necessity of the nuclear option.

Second, Israeli leaders present their country's supposed nuclear capability as a deterrent helping to stabilize the Middle East. They argue that Israel's presumed nuclear capability forces its adversaries to accept that Israel is here to stay. Israel's conventional weapon superiority and its nuclear arsenal make it an indispensable part of the Middle Eastern landscape. Israeli conventional and unconventional might, the argument goes, has forced Arabs to negotiate and reduced incentives for all-out war.

Third, Israel has aimed to monopolize the nuclear option, always seeking to deny its adversaries such capability. To achieve this end, Israel has employed diplomatic and military pressure against potential nuclear proliferators. The successful raid in 1981 on Iraq's nuclear facilities made Israel the first nation to launch a preemptive strike against an adversary's nuclear reactor, and the successful 2007 attack on Syria's nuclear plant was in line with this doctrine.

Fourth, Israel has hesitated to fully endorse the global non-proliferation regime. It has never placed its Dimona nuclear facility under the IAEA safeguards, nor has it since 1970 allowed any other type of inspection visits to that site. Israel has not signed the NPT or the Biological and Toxin Weapons Convention (BTWC); it has signed but not ratified the Chemical Weapons Convention (CWC) and the Comprehensive Test Ban Treaty. Israeli analysts argue nevertheless that Israel abides by the global norms and rules of the global non-proliferation regime. Gerald M. Steinberg claims that unlike Pakistan, whose chief nuclear scientist Abdul Qadeer Khan provided nuclear technology to several potential proliferators, Israel has refused to share its nuclear expertise with other regimes (except the South African apartheid regime).[26]

Fifth, Israeli leaders have repeatedly made a comprehensive peace between Israel and all Arab states and Iran a prerequisite to joining a NWFZ. Israel, they insist, would only cede its nuclear option if all its neighbors recognized and engaged in diplomatic and commercial ties with it. In other words, peace treaties would not be sufficient. Israelis require complete normalization of relations to ensure full acceptance from their neighbors.

These five characteristics suggest three conclusions. Israel is highly unlikely to endorse a NWFZ in the Middle East. The few statements Israeli leaders

[26] Gerald M. Steinberg, "The International Atomic Energy Agency and Israel: A Realistic Agenda," *Jerusalem Issue Brief*, Vol. 3, No. 27 (July 2004) at http://www.jcpa.org/brief/brief3–27.htm (last accessed August 10, 2004).

have made regarding such a zone indicate that they link their national survival to their nuclear weapons capability. Moreover, the refusal by Israel, India and Pakistan to join the NPT has prompted some analysts to suggest special arrangements to accommodate these three non-signatory states. Avner Cohen and Thomas Graham, for example, propose a form of NPT associate membership.[27] Others have called on Israel to follow the path taken by South Africa and relinquish its nuclear weapons.[28] Israel is unlikely to accept either proposal. Finally, the presumed Israeli nuclear capability and an Arab and Iranian lack thereof are considered a major reason for the proliferation of chemical and biological weapons in the region. Unable to achieve nuclear capability, Israel's adversaries have sought other WMD as second-best deterrents.

The Arab and Iranian approach

Although there is no single united Arab and Iranian approach to a NWFZ, Iran and most Arab states share similar sentiments. They do not see the Israeli nuclear arsenal as a "weapon of last resort" or "insurance policy" ensuring Israel's survival. Military asymmetry, and particularly Jerusalem's nuclear capability, is seen in Tehran and most Arab capitals as enforcing the occupation of Palestinian and other Arab territories and as a "primary threat to regional security and a factor of instability."[29] The fact that Israel is the region's only presumed nuclear power underscores and feeds a sense of Arab and Iranian technological and military inferiority. Arab countries vowed to drop out of the NPT if Israel ever officially acknowledges it has nuclear weapons.[30]

Iranian and Arab governments accuse Western powers of applying a double standard regarding Middle Eastern nuclear proliferation. From the Arab and Iranian perspective, the United States and major European powers have allowed—even assisted—Israel in acquiring nuclear weapons but have strongly

[27] Avner Cohen and Thomas Graham, Jr., "An NPT for Non-Members," *Bulletin of the Atomic Scientists*, Vol. 60, No. 3 (Chicago, IL: July 2004): pp. 40–4.

[28] Yossi Melman, "ElBaradei calls on Israel to give up nukes," *Haaretz* (Tel Aviv: December 12, 2003).

[29] Sami G. Hajjar, "Regional Perspectives on the Causes of Proliferation of Weapons of Mass Destruction in the Middle East," *Comparative Strategy*, Vol. 19, No. 1 (January 2000): p. 40.

[30] "Arab League Vows to Drop out of NPT if Israel Admits it Has Nuclear Weapons," *Associated Press*, at http://www.haaretz.com/hasen/objects/pages/printarticleen.jhtml?itemNo=961275 (last accessed March 5, 2008).

resisted Arab and Iranian attempts to develop similar capability. Many Arab officials have argued that so long as Israel maintains its nuclear option, Iran and other regional powers will have incentives to seek a similar capability. Accordingly, the most effective way to deal with Iran's nuclear ambition, some Arabs argue, is to pressure Israel to dismantle its nuclear weapons and join the NPT.

Arab states in the Gulf region took a number of initiatives to pave the way for the establishment of a NWFZ. In 2004 analysts at the Gulf Research Center, based in Dubai, explored the possibility of establishing a NWFZ covering the nine states of the Gulf region: Bahrain, Iran, Iraq, Kuwait, Oman, Qatar, Saudi Arabia, United Arab Emirates and Yemen. Representatives from several Gulf states discussed this concept of a sub-regional zone. The proposal was presented as a tactical step toward a broader NWFZ in the Middle East.[31] In 2008, the Bahraini Foreign Minister Sheikh Khald bin Ahmed Al-Khalifa called for the establishment of a regional organization that includes Israel, Iran, Turkey and Arab states, in order to discuss the issue of nuclear weapons.[32]

Egypt, with strong justification, considers itself the leader of the Arab world. Cairo led the Arab side in four wars against Israel. Thus, at least in the 1950s, 1960s, and early 1970s Egypt had incentives to engage in a nuclear arms race with Israel. However, the country's leaders have never allocated the necessary resources to build such capabilities. Instead, they have chosen to use diplomacy to convince or compel their adversaries to abide by the norms of non-proliferation regime. These efforts have largely failed with regard to Israel. The intense confrontation over Iran's nuclear program since the early 2000s has further fueled the Egyptian efforts to address the proliferation issue in the Middle East.

For a long time Egypt has adopted a clear two-fold position on the non-proliferation regime. First, for almost four decades Cairo has been calling for making the entire Middle East a nuclear weapons free zone, and it later expanded this call to include all weapons of mass destruction. Second, although Egypt signed and ratified the NPT, it has refrained from fully accepting commitments in other major agreements such as the CWC, the BTWC, the Comprehensive Test Ban Treaty (CTBT), the Fossil Material Cutoff Treaty (FMCT), and the Additional Protocol. According to Ezzedine

31 Mustafa Alani, "The Gulf NW and WMD Free Zone: A Track II Initiative," *International Relations*, Vol. 22, No. 3 (September 2008): pp. 358–62.

32 Yoav Stern, "Bahraini FM calls for Mideast Forum that includes Israel, Iran," *Haaretz* (Tel Aviv: October 1, 2008).

Ramzy, Egypt's Deputy Foreign Minister, Cairo has conditioned its acceptance of additional commitments on Israel's joining the NPT.[33]

The controversy over Iran's nuclear program is important to Egypt in an indirect way. The Islamic Republic of Iran does not pose a national security threat to Egypt. Unlike Israel, Iran has never gone to war with Egypt and has never occupied any part of Egyptian territory. Rather, Egypt has a two-fold concern about Iran's nuclear program. First, a nuclear Iran might become more assertive in pursuing its regional agenda. Former President Mubarak, along with other Arab leaders, was very suspicious of the Iranian policy. The post-Mubarak government in Cairo has expressed willingness to open a new page with Iran and seems willing to normalize relations with Tehran. Second, Iran's nuclear program underscores Egypt's inferior status regarding nuclear arms and nuclear technology in general. Egypt has never comfortably accepted that Israel has a monopoly of nuclear weapons in the Middle East. Iran with a nuclear weapons capability would further highlight Egypt's lagging behind in this race and nuclear technology in general.

Against this background, Egyptian officials have struggled to articulate a coherent stance. Former President Mubarak rejected a potential expansion of a US nuclear umbrella to include the Gulf states and Egypt. Mubarak stressed that such an umbrella "would imply accepting foreign troops and experts on our land and an implicit acceptance of regional nuclear powers. We do not accept that."[34] Answering a question on whether Egypt would go nuclear if Iran does, Nabil Fahmy, Egypt's former Ambassador to the United States (1999–2008) argued that it is not the only option. Egypt, Fahmy stressed, can pursue its national security interests by "taking political measures, balancing with other weapon systems, by limiting commitments to agreements and by trying to have a symmetrical response."[35]

Since Iran does not have nuclear weapons capability, these Egyptian options are speculative. Rather, three concrete characters can be identified in

[33] "Algeria, Emirates Plan Nonproliferation-Friendly Nuclear Programs; Egypt Keeps Fuel Cycle Options Open, Rejects Expanded IAEA Monitoring," *WMD Insights*, at http://www.wmdinsights.com/125/125_MEI_AlericaEmirates.htm (last accessed June 13, 2008).

[34] Fareed Mahdy, "Egypt Rejects U.S. Nuclear Umbrella," International Press Service, at http://ipsnews.net/print.asp?idnews=48156 (last accessed August 25, 2010).

[35] Peter Crail and Miles A. Pomper, "The Middle East and Nonproliferation: An Interview with Nabil Fahmy, Egypt's Ambassador to the United States," *Arms Control Association*, at http://www.armscontrol.org/node/3303/print (last accessed September 3, 2008).

Cairo's stance on Iran's nuclear program. First, Egyptian officials have repeatedly confirmed that like other signatories of the NPT, Iran has the right to benefit from the peaceful use of nuclear energy, provided that it proves its program is for peaceful uses. Second, Iran's nuclear dossier needs to be dealt with politically, not through military action. Cairo believes that a military action against Iran would further destabilize the region. Third, in order to avoid accusations of double standard, a comprehensive approach should be adopted that includes both Iran and Israel and leads to the establishment of a nuclear weapons free zone in the entire Middle East. It is not clear whether Egypt has the diplomatic and political leverage to convince or pressure other players to adopt its stance. Meanwhile, like other Middle Eastern states, Egypt has embarked on an ambitious program to expand its civilian nuclear power infrastructure.

Nuclear power and nuclear weapons

Since the early 2000s a large number of Arab countries, and many others all over the world, have expressed interest in establishing and developing nuclear programs to generate electricity, among other civilian uses. As in other regions, environmental, financial and strategic concerns are driving nuclear interest in the Arab world. Due to heavy oil subsidies, many Arab cities are among the most polluted in the world. Replacing oil by nuclear power in generating electricity and for water desalination would release more oil for export and increase national revenues. Finally, Arabs' interest in nuclear power can be seen as a potential hedging strategy against Iran's nuclear program. The relative weight of each drive varies from one Arab country to another. In addition, how they will address the uncertainties surrounding nuclear power is yet to be spelled out. What is certain is that the growing interest in nuclear power in the Arab world has the potential to drastically alter the security and strategic landscape in the entire Middle East and has significant implications worldwide.

The link between civilian nuclear power and nuclear weapons has been a serious unresolved worry throughout the nuclear age. The expansion of civilian nuclear power is likely to contribute to the dissemination of expertise, technology and material that would be useful in launching a nuclear weapons program. Stated differently, the connection between civilian and military applications of nuclear power seems inevitable. The technologies—uranium enrichment and plutonium reprocessing—provide the foundation for both nuclear power and nuclear weapons. This is dubbed the dual-use dilemma. Specifically, the challenge is that a potential spread of nuclear fuel cycle technologies, especially technologies for uranium enrichment and for reprocessing spent fuel to separate plutonium, poses a serious concern to the nuclear proliferation regime because enrichment and reprocessing capabilities give states the capability to produce fissile materials for weapons.[36] A commercial enrichment plant producing low-enrichment uranium (LEU) could be reconfigured to produce high-enriched uranium (HEU).

The growing global interest in nuclear power has focused new attention on nuclear fuel production. The uranium enrichment market is currently dominated by a small number of large global companies. This concentration has underscored the vulnerability of states with no national enrichment facilities and fueled their suspicion that external supplies of enriched uranium may become associated with political pressure. The Iran-EURODIF affair is a case in point. In the 1970s the Shah of Iran loaned more than US$1 billion to support the construction of the EURODIF diffusion enrichment plant in France. This gave Iran a 10 percent share in the consortium and made it entitled to 10 percent of the uranium enriched by EURODIF. With the 1979 revolution in Iran EURODIF refused to deliver the LEU to the Islamic Republic.

This uncertainty regarding the supply of nuclear fuel has contributed to the controversy over the adequacy of the global regime to find the "appropriate" balance between the full utilization of nuclear power for peaceful applications and the prevention of nuclear weapons proliferation. In 1957 the international community established the IAEA with the mandate to pursue such a balance, reassuring that community that civil nuclear programs are not contributing to weapons acquisition. In the intervening decades, the role of the Agency has grown, evolving in response to Member States' needs. After the uncovering of Iraq's clandestine nuclear weapons program in the 1990s, the IAEA's mandate was further strengthened by introducing the Additional Protocol which gave the Agency more inspecting power.

The NPT is the key foundation of the global non-proliferation order. The Treaty is based on a balance across three pillars—non-proliferation, disarmament, and peaceful uses of nuclear energy. These three pillars are

[36] Anatoly S. Diyakov, "The Nuclear Renaissance and Preventing the Spread of Enrichment and Reprocessing Technologies: A Russian View," *Daedalus*, Vol. 139, No. 1 (Winter 2010: p. 119.

integrally linked and the achievements in each area are likely to require progress in the others. The non-nuclear states insist on the priority of the second and third pillars, while the nuclear states emphasize the importance of non-proliferation.

The nuclear weapons states emphasize the objectives of preventing the further spread of sensitive nuclear technologies and material. They argue that the more states acquire their own national enrichment of reprocessing capability, the more worrisome the nuclear future will be. On the other side, the non-nuclear weapons states insist that the NPT incorporated the principle that states have an "inalienable right" to develop nuclear energy for peaceful purposes. Article IV states: "Nothing in this Treaty shall be interpreted as affecting the inalienable right of all the parties to the Treaty to develop research, produce and use of nuclear energy for peaceful purposes."[37] They further argue that it is a diversion to contend that the way to ensure non-proliferation is by limiting technological rights for states that are in compliance with their treaty obligations.

This ongoing controversy raises two largely unanswered questions. First, will the non-proliferation regime be adequate in a world where there is more nuclear knowledge and technology spread across more states?[38] And second, does every country need its own nuclear fuel cycle? Or would it be more economical, with minimal risks of proliferation and an effective verification system, to rely on multilateral fuel banks?[39]

In order to solve this non-proliferation conundrum, several proposals have been considered since the dawn of the nuclear age. Their common goal is to prevent an increase in the number of states that would be capable of producing usable in weapons nuclear material. Most of them have sought to reach this goal by proposing that international entities take charge of uranium enrichment or plutonium reprocessing instead of national authorities. In other words, the internationalization of the nuclear fuel cycle is their underlying theme. Most of these proposals are in general considered technically feasible,

[37] The text of the Treaty is available at http://www.un.org/Depts/dda/WMD/Treaty (last accessed December 24, 2010).

[38] Steven E. Miller and Scott D. Dagan, "Nuclear Power without Nuclear Proliferation?", *Daedalus*, Vol. 138, No. 4 (Fall 2009): p. 14.

[39] Mohamed I. Shaker, "Nuclear Power in the Arab World and the Regionalization of the Nuclear Fuel Cycle: An Egyptian Perspective," *Daedalus*, Vol. 139, No. 1 (Winter 2010): p. 93.

but lack political consensus and commitment, particularly among states aspiring to develop and acquire nuclear power.

In the late 2000s the US and the UAE signed an agreement on peaceful nuclear cooperation under which the UAE made a commitment not to enrich uranium or reprocess spent fuel. This is the first time that a cooperating partner of the US has made a legal commitment in a peaceful nuclear cooperation agreement to forgo enrichment and reprocessing. Washington seeks to establish the US-UAE agreement as the model for other states in the Middle East and elsewhere. It is unlikely that other Arab countries will follow this model, particularly if Iran is allowed to keep enriching uranium. In short, the evolution of Iran's nuclear program and the international community's response are likely to influence the development of nuclear programs in the Arab states on the Gulf and the broader Middle East. Finally, Israel is certain to be concerned about the growing interest in nuclear power in the Arab world and Iran. How Israel would react to the construction and operation of nuclear reactors in the Middle East is yet to be seen.

To sum up, given the strong connection between peaceful and military uses of nuclear power, the growing interest in this source of energy (despite Japan's nuclear disaster in early 2011) has raised security alarms and added more uncertainty to the political-strategic landscape in the Middle East.

A nuclear weapons free zone in the Middle East: the way forward

Several conclusions can be drawn from the discussion of the five established NWFZs and the unsuccessful efforts to create one in the Middle East. First, experience all over the world suggests that a certain level of peace and accommodation is essential to reach a consensus on NWFZ. Peace and non-proliferation go hand in hand and reinforce one another. Given the deep-rooted conflicts in the Middle East, a NWFZ would require considerable changes in the political landscape, including a settlement of the Arab-Israeli conflict, stability in the Persian Gulf, and a rapprochement between Israel and major Gulf states.

Second, settling these conflicts and creating the right environment for a comprehensive and durable peace will take a long time. In the short and medium terms, Middle Eastern states need to implement confidence-building and arms-control measures that would reduce the underlying causes of violence and war. They need to engage in mutually beneficial economic and cultural relations. A slowdown or containment of conventional arms races would enhances prospects for regional peace. According to the Stockholm International Peace Research Institute (SIPRI) Middle Eastern states accounted for 17 percent of international transfers of major conventional weapons in the period 2006–10, down from 21 percent in 2001–5. In the second half of the decade, 23 percent of all major arms transfers to the Middle East went to the UAE, 14 percent to Israel and 13 percent to Egypt.[40]

Third, foreign powers play a significant role in establishing and maintaining NWFZs. Part of their role is providing the so-called negative security assurance and positive security assurance. The former is a declaration by a nuclear weapons state not to use nuclear weapons against a non-nuclear weapons state; the latter is a pledge to aid a non-nuclear weapons state if it is the victim of a nuclear attack. Occasionally, some global powers have maintained a level of ambiguity regarding these commitments. What has been over three decades of deep hostility between Washington and Tehran pose a major hurdle to the prospects of establishing a NWFZ in the Middle East.

Finally, many Israelis argue that their nuclear arsenal is their last line of defense and they should not give up such a capability unless Iran and all Arab states completely accept Israel. On the other side, many Arabs and Iranians insist that they cannot negotiate peace with Israel without prior nuclear disarmament. Iran's nuclear program and the recent uprisings in several Arab countries add more uncertainties.

[40] "India World's Largest Arms Importer According to New SIPRI Data on International Arms Transfers," paper at Stockholm International Peace Research Institute, at http://www.sipri.org (last accessed March 14, 2011).

Index